ψ

Physical Truth

A Mathematical Philosophy

Bruce Rout

FOXCAMP
D e s i g n

Cover by David Vink

ISBN: 978-0-9812096-2-3

For Tara

Comments on Physical Truth

I looked over your paper you sent me. It is a fun read in the vein of 'Zen and the art of Motorcycle Maintenance.'

- Dr. Keith Promislow
Michigan State University

Very cute. However, I believe there is a cosine missing on page five.

- Allen Murray
Queen Elizabeth Golden Jubilee Gold Medal Winner
Early Childhood Education

I enjoyed reading what you sent. You have an engaging, informal style laced with plenty of humour. I loved the "Superman's Beard" story which was new to me.

- Dr. Werner Israel, (OC)
University of Victoria
Victoria, B.C.
Canada

Contents

A Game

Long, long ago, the Canadian government issued its one dollar currency on paper, rather than as coins. The bills had serial numbers that people used in order to play Bullshit Poker. The digits of the serial numbers on each bill were used to make an equivalent of a hand of cards in a game of poker. Suppose, for example, you had a bill with the serial number 1114267674. You might start out saying "1 ace." The other player might come back, after looking at her bill, with "2 aces." Then you know she has at least one number 1 on her bill, unless she is already bluffing. If you were in a hurry you might jump to "4 aces" in the hope that she didn't have any more than one 1 on her bill. Both sides continue bidding higher and higher poker hands until one of the players calls "bullshit" on the other player. The two bills are examined and if the last player to bid cannot combine the numbers on both bills to fulfill their hand, they lose. If the numbers on both bills substantiate the last bid, they win. The winner would walk away with both bills. Mostly it was a time killing game, played in the waiting room of a bus or train station, or in the bar car on a train until a CN employee caught you and told you there's no gambling on the train.

Preface – A Personal Introduction

What is Truth? Does it exist? How can we know it and recognize it? Should we follow it and adhere to it? Does it matter?

The Truth exists. The Truth is real. The Truth is so real that the entire Universe is made up of the stuff.

Our culture, the culture of the western world, the culture of northern Europe, is very, very recent. Our culture goes back only about two to three thousand years – it is just at its bare beginnings and our recent way of life is but a few centuries old. And yet, even we, of a materialistic and self indulgent society, have our cultural roots in the search for Truth.

The Greeks thought Truth to be a supreme virtue. Vikings and Celts thought Fate could be changed through acts of great courage. The middle ages saw the birth of our legal system based on the belief that Truth was so powerful that a honest man could not possibly lose in a trial by combat or be abandoned by the Source of Goodness in an ordeal.

My journey has come to an end. I was walking along the banks of a dried up river in Southern Alberta. I was thinking of Truth and the many truths that follow from it. "How am I going to explain this?" I thought. I looked down and saw stones in the river bed, many of them. I saw one in particular and reached down and picked it up. "Ah," I thought, "this will do nicely. A perfect example of Truth." I held the stone with a degree of gratitude. It is so remarkably simple.

I thought back. My search for Truth began a very long time ago. I was rather young, I guess 10 or so in the late 1950s, and I read the life of Socrates from a wonderful book for boys titled, "Socrates". I really admired the ancient philosopher for laying down his life for what he believed in, esoteric though it was, and grasping the crown of martyrdom when it was offered to him. He said he was an old man anyway and, what the heck, may as well go out defending what he believed in. Cool.

Obviously, I didn't get out much. Actually it was because of my father who is very old now and is about to step into the next world. I am also getting on but still have a few decades to wait. Dad was a high school teacher and to say he was pragmatic would be a gross understatement. I was ordered to my room and forced to stay there for at least four hours every night to "study" from the time I was in grade four. There is not much homework or study to do in grade four, but that was the way it was. There were no electronic games; they hadn't been invented yet. As a matter of fact, TV had just been invented only a few years earlier. All I had to keep me occupied was a copy of the Elements by Barnard and Child, a compass, with which to draw circles, a straight edge and an eraser to poke holes in with the sharp end of the compass. I could do every construction in the Elements by heart and had set out to trisect an angle just to keep my mind alive. I got to be very good at Euclidean Geometry. Sometimes I snuck down into my Dad's study and read old books. Socrates was one of them.

This journey came to a major turning point in grade nine. I was taking physics and hadn't a clue what it was about. It made no sense to me at all. Breaking down, I asked my Dad for help with some of the problems while we were at the supper table.

"What do you need help in?" he asked, "Is it your math?"

"No Dad, it's physics," I replied.

With the mention of the word "physics" my father froze and I could see the fear of God Almighty cross his face. He was terrified.

"No," he said, "I can't help you with physics. You'll have to ask your prof."

He lowered his head to resume eating, avoiding my look. I had to take the shot. My father had never shown fear of anything before. I was beginning to see the locked door of my bedroom where I studied every night begin to crack open and there was light on the other side.

"No Dad," I said, "It's just ... PHYSICS!!!!" to see what would happen.

My dad jumped back in his chair wide eyed in terror.

"No," he said hurriedly, "I can't help you. You have to ask your teachers." He then jumped up from the table and quickly ran out of the kitchen. His supper was left barely touched on the table.

Needless to say, the next day I was down at the local library begging the librarian to show me everything there was about physics. All they had was the Encyclopedia Britannica. I looked under "P" for physics and still remember the opening statement on the subject from the most prestigious encyclopedia in the world at that time.

Physics is in a crisis. No one can agree to its fundamental principles or even to what the subject is about. This terrible circumstance is the result of the fact that no one can understand or comprehend a certain symbol. This symbol is:

$$\psi$$

And it went on. It talked about probability but said the author of this terrible symbol, Schrödinger, declared emphatically that it could not be probability. It was all very important. It all came from this formula:

$$i\hbar\frac{\partial\psi}{\partial t} = -\frac{\hbar^2}{2m}\nabla^2\psi + V\psi$$

I stared in disbelief. I mean, I'm 13 years old, my mother has just been killed in a car accident, I'm still locked in my room and I cannot get out unless I figure out what this stupid symbol means. And all of the best minds in science have been unable to do so, even the guy who came up with the thing in the first place. This was going to take a lot of time. It did.

I took some quantum mechanics and atomic physics in 1975 in university but then switched to astrophysics. I was studying quasars. These things are 40 times brighter than an entire galaxy, have a period of luminosity of about a month and a half, (which means they can't be bigger than a light-month-and-a-half or so) and have huge red shifts. Quite a few have huge blue shifts. I had begun with pencil and paper, as is my lot, and was mapping out the positions of the quasars on a grid. I placed an arrow pointing upwards from the position of each quasar. The length of the arrow was the length of the red shift. If the quasar was blue shifted, the arrow pointed down. It didn't take too long to realize that many quasars come in pairs. I checked to see what was between each pair and found a galaxy. The gravitational field of the galaxy was causing a gravitational lens effect. A distant quasar behind the galaxy had been "split" into two images on either side of the galaxy by light curving around either side of it. There seemed to be a lot of them. I ran about the campus and tried to get a paper together and to publish. I was only an undergraduate. I didn't know how to publish a paper. Three months later Physics Today published an article on the gravitational lens effect of galaxies on distant quasars. I was pissed.

There is not only a lens effect, there is also a prism effect. There is chromatic aberration in the gravitational lens and the gravitational field is causing distant light to red shift as it passes through gravitational fields. Later that year we found out that the universe is not expanding. There was no big bang. It was going to take a long time to prove it. It did.

Sometimes you get lucky. I have been extremely fortunate in my life to have met and worked with exceptionally nice people who are experts in their fields. Dr. Doug Hube taught me astronomy. Doug was/is the top observational astronomer in the world. I worked with and became very good friends with Dr. Gary Margrave, of geophysics fame, with Dr. Werner Israel, the top expert in the world regarding black holes and gravitation, and my dear friend, Dr. Keith Promislow, number one in applied mathematics. All led me to the discovery of Rout's Law: **The only way to win at mathematics ... is to cheat.**

I worked with Gary doing research into earthquake prediction in the late 70s. He introduced me to tensor calculus and showed me a proposed Unified Field Theory from Misner, Thorne and Wheeler. I went to Werner Israel's office to ask his opinion on the Unified Field Theory. I wrote it on the board. Dr. Israel got very angry with me and said it was crap. He erased it off the board and took about ten minutes or so to work out the Unified Field Theory.

"There," he said, "that is the Unified Field Theory. I don't care what you do. Write it down, memorize it or figure it out. I don't ever want to have to do that again." Of course, I forgot it and had to call him back thirty years later. He was not angry. He was delighted.

Shortly after my first meeting with Dr. Israel, I worked at the Alberta Research Council on recovering oil from the Athabasca tar sands. There I met Dr. Teddy Schmidt and got very heavily into trying to find general solutions to non-linear differential equations. In particular, the free boundary problem. It took 23 years to find the solution. We did it using an approach we called the general operator.

Keith, at Simon Fraser University, showed me a perturbation technique I could use to set up differential equations and, using the general operator, we cracked the free boundary problem. It was time to return to the Einstein Field Equations.

After a few years, I got to play with a radio telescope at the Rothney Observatory in southern Alberta, thanks to Fred Babott, the observatory's head technician. I found a way to measure the redshift of light passing through a gravitational field. I could relate the discovery to the field equations. I could solve Schrödinger's Equation. I knew what ψ was. I called Werner Israel, after 30 years, at the University of Victoria where he was then working.

"Hello Dr. Israel," I said, "You don't remember me but I was a student of yours thrity years ago."

"No, I'm sorry but I do not remember you," he replied.

"I wouldn't doubt that," I said, "However, you gave me a little problem to try and solve way back then and I believe I have something. I wanted to tell you about it."

"Oh yes," he chuckled, "what was the problem?"

"The Unified Field Theory."

"Oh," he said, "The Unified Field Theory... **Physics is in a great crisis. Nothing makes any sense. There are now an infinite number of universes with an infinite number of outcomes to every event. It is a terrible, terrible end to physics.***"*

"Yes, I know," I said quietly, "I think I might have something. I can relate Schrödinger's Equation to the Einstein Field Equations and it looks like it all makes sense."

Silence ... "Well," said Dr. Israel, "If what you are saying is true, it would be a huge breakthrough in field of physics."

More silence ... (with humility) "Yes Dr. Israel, that is correct."

Quite a bit of silence ... and Dr. Israel broke out with a good and hearty laugh.

Shortly after we spent an afternoon going over the Unified Field Theory and an approach to deriving quantum mechanics from classical theory.

And a couple of months later, I was walking by a dry river bed and saw this stone ...

Abstract

A proposal for a unified field theory is presented. A validation of the axiomatic approach to establish mathematical truths is given with discussions on Gödel and Russell. An existence criterion is developed. A general solution to the time-dependent Schrödinger equation with boundary conditions is found to derive the Heisenberg uncertainty formulae, an insight to quantum mechanical photonic energy and a possible calculation of spin-states. A general relativity/quantum mechanical interaction between a photon and the Sun's gravitational field is examined to determine the degree of red shifting of light passing through a gravitational field. The Einstein field equations, complete with an arrangement of Faraday tensors, are presented with suggestions to determine the energy of a photon from Einstein's and Maxwell's equations. Schrödinger's equation coupled with the Einstein field equations and Maxwell's equations is used to derive a postulated foundation for string theory. A distance measure to NGC 3198 is presented and verified. This distance measure is applied to galaxies in the Southern hemisphere invalidating the theory of an expanding universe. The General Theory of Relativity is applied to a rotating reference frame to determine the effects of negative curvature and derive Roxy's Ruler. Cepheid variable measurements are validated. A closed form analytic model is presented to describe the shape, structure and luminosity profile of spiral galaxies. The flat velocity curves of galaxies is explained. An entropy reversal process is found to be possible at the centre of galaxies. High energy jets of fundamental particles are shown to be the product of entropy reversal in the polar regions of very fast spinning black holes. Extreme energies in the form of cosmic and gamma rays are explained by the model. The information paradox is resolved. A resultant philosophy is presented.

Table of Symbols

$\rightarrow$	means	it follows that, ie. $A \rightarrow B$ means if statement A is true then statement B is true.
$\leftrightarrow$	"	if and only if, if A then B *and* if B then A.
$\therefore$	"	therefore.
$/$	"	the compliment of a set. Everything that is not in the set.
$\exists$	"	there exists.
$\forall$	"	for all.
$\in$	"	is a member of.
$\notin$	"	is not a member of.
$\{\dots\}$	"	a set defined by its members.
$\mid$	"	such that.
$\neg$	"	not. The negation of the statement.
$/$	"	if through a symbol, then the negative of the symbol.
$\vee$	"	or.
$\wedge$	"	and.
$\dots$	"	the words "and so on".

A Very Brief Introduction to Peano Arithmetic

We reduce a statement to a letter. For example, let's make the statement, "It is raining", be represented by the letter A. And let's make the statement, "The ground is wet", be represented by the letter B. And when we say, "If it is raining then the ground is wet" is represented by:

$$A \rightarrow B. \tag{1}$$

We assume the word "if" is in front of the equation (1). So the arrow, $\rightarrow$, forms an if statement between A and B. We assume the statement $A \rightarrow B$ is true unless we prove it otherwise. We ask you to consider that in this case there is no cheating. There is no cover over the ground and we are playing fair. If it is raining then the ground is wet. So far so good.

Now let us use the "not" symbol to create a different statement, such as:

$$A \rightarrow \neg B \tag{2}$$

which is false if the first statement is true. Because we are saying that if it is raining the ground is wet means that if it is raining the ground cannot be dry (not wet), it must be wet. So we have a true and a false statement. The first statement, $A \rightarrow B$, is our hypothesis. We assume it to be true. Continuing on:

$$\neg A \rightarrow B \tag{3}$$

says that if it is not raining the ground is wet. This could be perfectly true. You may have thrown a bucket of water on it when I wasn't looking. So, if we agree that if $A \rightarrow B$ is true then $\neg A \rightarrow B$ could also be true but $A \rightarrow \neg B$ must be false. So far so good.

Let us look at the converse which is:

$$B \rightarrow A \tag{4}$$

which says that if the ground is wet, then it is raining. Well, we see that may be true, but again you may have thrown a bucket of water on the ground to trick me. So it is not necessarily true. Continuing on:

$$\neg B \rightarrow A \tag{5}$$

Says that if the ground is not wet then it must be raining and we have ruled that out, so that must be false. Fair enough. However, consider this:

$$\neg B \rightarrow \neg A \tag{6}$$

means that if the ground is not wet, then it is not raining. And that is definitely true. If you are outside and someone asks you if it is raining, you put out your hand to see if it gets wet. If it doesn't, then you know it is not raining. If it is wet, then it could be raining or I could have thrown a bucket of water on you to get you back for tricking me.

$$A \to B \tag{7}$$

is the hypothesis and

$$\neg B \to \neg B \tag{8}$$

is called the negative converse to the hypothesis. This was our first lesson in Peano Arithmetic which was taught to us in grade 9.

We now have a theorem:

$$(A \to B) \leftrightarrow (\neg B \to \neg A) \tag{9}$$

and you can put that into words as in our example and try to see if it is true.

Chapter 1

A Little Theorem

To begin our discussion we will explore existence. We will look at numbers and what a number is. We are starting on a path of looking at mathematical truths and then expand that path to look at the world of physics and the truths that lie within. As we develop our discussion we will move step by step from established and well-understood concepts and facts. We hope to present an argument with some degree of rigour. Note that because of Gödel's Incompleteness Theorem, [URL-8], we can propose a consistent and valid argument which is not complete in that it does not prove the axioms themselves. We present a discussion of axioms to attempt to construct as complete an argument as possible. We begin with the existence of a number.

Question: Does a number exist and we discover it or do we just make numbers up as we go along?

Theorem: Numbers exist.

Proof:

Definition: A number is the name of the set of all sets having the same magnitude as the number. The magnitude of a set, (or number of elements in the set), can be determined through a matching process of its members to a set of ordinals. A set of ordinals is a set whose members are in a determined order. An example of an ordered set is the set $\{1, 2, 3 \ldots\}$, [Russell & Whitehead (1910-13)], [URL-9].

Example: The number one is the name of the set of all sets having one element. The number two is the name of the set of all sets having two elements, and so on.

Definition: The null set, or $\emptyset$, is the set containing no elements.

Definition: The universal set, or $\mathcal{U}$, is the set containing all elements except itself, which will be demonstrated shortly.

It can be easily shown that:

$$\emptyset = \mathcal{U}'$$
$$\therefore \quad \mathcal{U} = \emptyset', \tag{1.1}$$

hence,

$$\emptyset = \{\}$$
$$\therefore \quad \emptyset \neq \{\emptyset\}$$
$$\therefore \quad \emptyset \notin \emptyset$$
$$\therefore \quad \mathcal{U} \notin \mathcal{U}. \tag{1.2}$$

Consider zero. By definition, zero is the name of all sets having no elements. There is only one set having no elements. Therefore:

$$0 = \{\emptyset\} \text{ by definition.} \tag{1.3}$$

Consider the solution set, S, which is the set of all solutions to all problems. Suppose we have a problem which we will call P. If there exists a solution to this particular problem P then its solution is a member of the set S. If no solution exists to the problem P then we say the solution is $\emptyset$. As a result, if there is no solution to P then $\emptyset$ is a member of S and if a solution does exist to P then $\emptyset$ is a member of $\mathcal{S}'$. We then have the well understood situation:

$$\exists \emptyset \rightarrow (\emptyset \in \mathcal{S}) \wedge (\emptyset \in \mathcal{S}'). \tag{1.4}$$

Although this appears as a contradiction, it is a fairly well known property of the null set. The null set is a member of all sets, except itself.

2

We have for existence properties, if $\mathcal{A}$ is proposed as a possible solution to a problem, using $\neg$ as "not":

$$
\begin{aligned}
\neg\exists\mathcal{A} &\quad\rightarrow\quad \exists\emptyset \text{ and} \\
\exists\mathcal{A} &\quad\rightarrow\quad \exists\mathcal{A}' \vee \exists\emptyset \text{ or else} \\
\exists\mathcal{A}' &\quad\rightarrow\quad \exists\emptyset.
\end{aligned}
\tag{1.5}
$$

We conclude that if a set does not exist, independently of whether or not it is the solution to a postulate, (or problem P), then the empty set exists.

$$
\neg\exists\mathcal{A} \rightarrow \exists\emptyset, \forall\mathcal{A} \in \mathcal{U}
\tag{1.6}
$$

Note that the interior of the null set is nothing, but the null set itself exists, [Halmos (1974)], [Jech (2003)]. This can be shown by:

$$
\exists\emptyset \rightarrow \exists\emptyset, \text{ if we assume the null set exists}
\tag{1.7}
$$

and

$$
\neg\exists\emptyset \rightarrow \exists\emptyset, \text{ if we assume the null set does not exist.}
\tag{1.8}
$$

We see that whether the null set exists or not, it exists. Whether the interior of the null set exists or not is a matter of philosophical argument and beyond the bounds of this discussion. Nevertheless, the null set itself exists. Therefore:

$$
\exists\emptyset \rightarrow \exists\{\emptyset\}
\tag{1.9}
$$

and

$$
\exists\{\emptyset\} \rightarrow \exists 0
\tag{1.10}
$$

and, of course:

$$
\begin{aligned}
\exists 0 &\quad\rightarrow\quad \exists\{0\} \\
&\therefore\quad \exists 1 \\
&\therefore\quad \exists\{0,1\} \\
&\therefore\quad \exists 2
\end{aligned}
\tag{1.11}
$$

and so on.

This, of course, leads to the interesting theorem:

$$
(\exists\emptyset \rightarrow \exists\emptyset) \leftrightarrow (\neg\exists\emptyset \rightarrow \exists\emptyset)
\tag{1.12}
$$

3

which is a contradiction but not a negation of the theorem since it is a property of the null set. We shall refer to (1.12) as Our Little Theorem.

Numbers therefore exist because the null set exists and the number 0 exists. We note that zero represents nothing and the non-existence of something counted, but zero itself exists. Similarly, whether or not the members of the null set exist, the null set itself exists.

Note that previously Our Little Theorem was presented as a part of the definition of the Null Set in order to bypass the consequences of Russell's Paradox. We show that the same result can be derived from a more direct definition of the Null Set and (1.12) is non-axiomatic.

This is an email from my son, Cam, after he read the above proof that numbers exist and after I had a little exchange with Dr. John Nash of Princeton University[1]:

Dad,

Zero is the set of nothingness and not nothingness itself, which as you described, is the contents of the set. The set itself can be counted, but its contents simply do not exist, by definition.

When considering negative or positive zero, we are dealing with one set and two operators. The operation conducted by the sign on an integer differentiates between two different sets, the set represented by the positive number and the set represented by its negative counterpart. Since there is only one null set, then the operating sign cannot be differentiating between two different sets and therefore zero has neither a positive nor negative sign.

However! The null set can be counted, which makes zero an integer and subjects it to integer operators such as a sign. Zero itself is neither positive nor negative because it has no counterpart, but the result of the operation of a sign

---

[1] I had been corresponding with Dr. John Nash, of 'Beautiful Mind' fame, over finding solutions to non-linear differential equations. He was still at Princeton at the time. I had asked if he had proven a theorem that all problems have solutions. It was important to the construction of a general differential operator. A general differential operator includes both linear and non-linear operators, a linear operator plus non-linear operator. Dr. Nash responded that he did not prove such a theorem and said not to believe anything that was attributed to him even though it may be published work with his name on it. He cited Fermat's Last Theorem as an example of an unsolvable problem: Given:

$$x^N + y^N = z^N \tag{1.13}$$

for any N, find x, y, and z where x, y, and z are integers to make the equation true. Fermat's Last Theorem said there were no solutions for $N > 2$. I wrote back that if $x, y, z = 0$ then there was a solution for all N. Fermat's Last Theorem stipulates x, y, and z must be positive integers. I never heard back from Dr. Nash. I think I may have pissed him off.

on zero is an integer, and since the result of the positive and negative operation involving a sign and a zero have different results.

(tongue goes in cheek here).

For example take $\lim_{x \to \infty} \left(\frac{1}{x} \right)$.

When $x = 0$ the result is undefined
When $x = +(0)$ the result is positive infinity.
When $x = -(0)$ the result is negative infinity.

Ok, so it's not "correct" arithmetic, but my point is that $+(0)$ and $-(0)$ are different. By your own definition of equality, when the operation of a sign on zero is replaceable by zero itself, then they are equal. However, this is not true in the above arithmetic so I will go ahead and say:

$$
\begin{aligned}
0 &= 0 \\
0 &\neq +(0) \\
0 &\neq -(0) \\
-(0) &\neq +(0)
\end{aligned}
$$

Since result of the operation of the positive sign on zero is different from the result of the operation of the negative sign on zero, then they both must exist and they are different from the number zero.

Conclusion:

1. *Zero is neither positive nor negative, but it is an integer*

2. *The positive and negative sign operators have different results when applied to zero and therefore the results of each operation are integers that exist.*

3. *The integer resulting from the operation of the positive sign on zero is in fact positive since it has a negative counterpart (unlike zero itself).*

I propose that the positive integer which results from the operation $+(0)$ does indeed solve Nash's problem.

- Cam

I wrote back:

Cam,

Personally, I'm going to have to think about this. Werner was very surprised to hear there was any difference between a positive and negative zero and I *haven't heard back from John yet. I like the countability idea. We say that, given $f(x) = \lim_{x \to \infty}(\frac{1}{x})$, if we approach $x = 0$ from either the left or right of the y axis, $f(x)$ appears in two separate regions which are an infinite distance apart. So, it makes a difference as to whether we approach the limit from either side, from the positive or negative. It appears we get two different positions for $f(x)$ at the same time and hence, we say $f(x)$ is undefined at $x = 0$. That being said, I don't see anything really wrong with your argument per se. I believe your saving grace is in using equivalency rather than equality. Your use of Aristotle's second statement is admirable.*

Chapter 2

Equality

Geometrically, if two triangles can be placed on top of one another such that they coincide, they are said to be congruent, [Barnard (1914)]. If so, then one triangle can replace the other.

If there are explicit conditions, an entity can be replaced by another. If conditions are considered as a boundary and entities meet these conditions, we say these entities are within this boundary and thereby form a space. All entities which can replace each other within this space are said to be equal under the restrictions of the conditions described by the boundary.

For example, let there be a ruler 10 cm. long. We may measure something that is 10 cm. in length using this ruler. However, let us say we also have two sticks, each 5 cm. in length. If we combine the lengths of the two sticks we have a length that can replace the 10 cm. ruler. Therefore, under the condition of "length", however we wish to define it, the two sticks are equal (in length) to the ruler. Nevertheless, under the condition of counting, or number, the ruler is not equal to the sticks. The number of ruler – one, does not replace the number of sticks – two. There is one ruler and two sticks. But under the conditions of length, the ruler and combination of two sticks are equal. In the space bounded by considerations of length, any combination of anything having lengths that can be replaced by each other are equal in length.

We see that in a bounded space so described, that if things can replace each other then they are equal. Furthermore, all things that can replace a particular entity within the specified boundary, can replace each other. From this we deduce that all things equal to the same thing are equal to each other. Equality requires condition.

Higher restrictions of the boundary condition leads to congruence and various definitions of congruence and resultant properties.

We see that this is nothing more than an axiom of Euclid. We define an axiom as an unprovable truth. A proof is the resultant derivation through defined and substantiated steps to a conclusion from a set of axioms. Considering the field of elements and operations, we say that a proof derives mathematical truths.

Chapter 3

Russell's Paradox

Russell's Paradox, [Russell (1902)], can be stated as:

$$(A \to B) \leftrightarrow (\neg A \to B) \qquad (3.1)$$

In words, this says that if A then B, if and only if, if not A then B. This is for any theorem or predicate. Which means that given any predicate, there is a contradiction and thereby no predicate can be true. We have an exception to this statement in our Little Theorem. Therefore, any theorem based on this paradox is disproved by exception.

Russell's Paradox is based on proper sets. A proper set is a set that is not a member of itself. The set of all proper sets is called the set $\mathcal{R}$. Since $\mathcal{R}$ is itself a proper set, it then contains itself and thereby results in a contradiction. This forms the basis of difficulties with the axiomatic approach in establishing mathematical proofs which have haunted mathematical logic since the beginnings of the 20th century. However, we see that a contradiction does not necessarily establish a falsehood. There exist entities of truth that can produce a contradiction yet do not invalidate the entity if it happens to be a property of the entity itself. The null set is one such entity concerning its existence.

There should be an obvious reason for this. Since we are using principles of Aristotelian logic in our deductions we turn to the *Categoriae* of Aristotle, [Aristotle (335 BC)], that makes three statements upon which set theory is based:

1. we can classify things,

2. there is a difference between a thing being classified and the classification itself,

3. and no classification may be a classification of itself.

From this is it easy to see that:

$$\mathcal{R} = \mathcal{U}$$
$$\mathcal{R}' = \mathcal{U}'$$
$$\mathcal{R}' = \emptyset$$
$$\emptyset \notin \emptyset$$
$$\mathcal{U} \notin \mathcal{U}$$
$$\mathcal{R} \notin \mathcal{R}.$$

(3.2)

From the above derivation we note that the statement that, "the set of proper sets must be a member of itself" does not follow. This is because the set of proper sets is the universal set and a property of the universal set is that it is not a member of itself. The universal set is the set of all elements, (which also includes sets; sets can be members of other sets but not of themselves), except itself. It is an exception to its own definition since it is the compliment of the null set. By definition, the null set is not a member of itself. Again, a contradiction does not necessarily denote falsehood. A contradiction alone is not sufficient to prove falsehood. If zero is a positive integer, Fermat's Last Theorem is disproven; if zero is not a positive integer, Fermat's Last Theorem stands.

Occam's Razor, [URL-6]

Occam's Razor is a well known dictum to enforce the reduction of assumptions in logical proofs. Basically, Occam's Razor states that assumptions must be kept to an absolute minimum. [1]

---

[1] If we see mathematical truths can be established some mention of the razor should be made and it's application to Russell. A postulated lemma of Occam's Razor is quite the opposite to conclusions drawn as a result of Russell's Paradox. Namely, what I refer to as Miller's Lemma: *Every problem has an infinite number of solutions but only one is the best and* **that** *is Occam's Razor.* This statement postulates that there is such a thing as a best solution out of many and that the best solution is given by the reduction of assumptions. It proposes that there is only one "best" solution. It may be an attribute of the human mind to be able to recognize the best solution or even what is meant by "best". From this line of thinking we believe that paradoxes pose contradictions, puzzles and riddles, but all can be resolved. The resolution may not answer the question asked by the paradox itself, but a resolution may trivialize the paradox.

If there are a number of solutions then the one with the least assumptions is the "best". This can be used to resolve paradoxes. An excellent example is "Superman's Beard", which is also known as the Barber Paradox.[2]

Barber Paradox

In Spain there is a city named Seville. There was a famous opera called The Barber of Seville by Rossini, near the beginning of the 1800s. Sometimes this paradox is confused with this opera. I do not know if Russell and Whitehead were thinking of Seville, Spain, or of a famous Rossini opera. The Barber Paradox is quite famous and sometimes attributed to Russell. In fact he attributes it to an unnamed person who suggested it to him, [Russel (1914-19)].

> That contradiction (Russell's paradox) is extremely interesting. You can modify its form; some forms of modification are valid and some are not. I once had a form suggested to me which was not valid, namely the question whether the barber shaves himself or not. You can define the barber as "one who shaves all those, and those only, who do not shave themselves". The question is, does the barber shave himself? In this form the contradiction is not very difficult to solve. But in our previous form I think it is clear that you can only get around it by observing that the whole question whether a class is or is not a member of itself is nonsense, i.e. that no class either is or is not a member of itself, and that it is not even true to say that, because the whole form of words is just noise without meaning.
>
> – Bertrand Russell, The Philosophy of Logical Atomism, [URL-2]

The paradox itself is as follows:

---

[2]Superman, [URL-7], is the famous comic hero from Krypton who is known as the man of steel. Superman was first published in 1938 during the Depression. He was bullet-proof, super strong, could fly and had extraordinary powers. The children reading the comic were invited by the comic staff to find logical reasons criticizing the existence of Superman and the stories in the comics. One child wrote in, noting that Superman was clean shaven but not bald. His hair grew. How could Superman shave if no metal on earth was strong enough to cut through his beard? The staff at Action Comics, which published Superman, was delighted with the question and a contest was started to see if anyone could figure out how Superman shaved. Further criticisms to disprove the existence of Superman were encouraged with a year's supply of comics for anyone who could stump the writing staff. Only the question of "how can Superman shave?" survived the attempts of the writing staff to fend off disproofs of Superman's existence or logic of various story lines. The question still remains.

There lives a very special barber in the town of Seville. This very special barber has a most unusual characteristic. This barber shaves every man who does not shave himself. The question is: Who shaves the barber?

If someone else shaves the barber then the barber would not be shaving himself. But the barber shaves *every* man who does not shave himself. Therefore, someone else cannot shave the barber.

If the barber shaves himself we note that the barber *must* shave every man who does *not* shave himself. Therefore the barber cannot shave himself.

So, who shaves the barber?

We re-examine the paradox with emphasis on assumptions. [3]

This barber shaves every *man* who does not shave *him*self. Who shaves the barber?

Resolution: Who cares? The barber is a woman.[4]

---

[3] At the time of this paradox, the *Principia*, Russell and various other philosophical discussions, there lived George Bernard Shaw. Shaw was a playwright in love with irony and a staunch socialist. His *Man and Superman* was a counter argument to Friedrich Nietzsche's philosophical ideas concerning the evolution of a "superman" or super race. The new Superman is woman. And the only razor sharp enough to shave Superman's Beard, is Occam's Razor.

[4] Wikipedia states this is a trivial solution to the paradox. More formally, the barber cannot be of the class that is being shaved. If the barber shaves every person who does not shave themselves, then the barber is not a person. It may be a machine, etc. If there is the insistence that the barber is shaving every class of the universal set, or no alternative from what is being shaved exists, then the original statement, "There is a very special barber in the town of Seville." is false and therefore has been proved to be false. In other words, it is a lie. Just as every statement included in this discussion is a lie, especially when you are told that you are being lied to. That, of course, is the biggest lie that could be told, that I am lying to you when I tell you I am lying to you. And that is no word of a lie either. The resolution of the Liar's Paradox is left as an exercise for the reader.

The Library Paradox, or List Paradox, is that the alphabetical list of all things beginning with a particular letter would have, under the L's, an entry that contains the "List of all things beginning with the letter L" and thereby contain itself. However, the listing does not contain itself, it contains the name of the list, not the list itself. In order to contain itself, it would require an infinite amount of paper, computer storage space, whatever, since the situation is self-recursive. In sum, we are stating that all paradoxes can be resolved.

Chapter 4

The Observer

Without going into all the references, there has been a great deal of discussion of the rôle of the observer in quantum mechanics. This comes from what is termed as the Copenhagen Interpretation, which, in very basic terms, says the ψ-state describes a probability of existence. Both Einstein and Schrödinger opposed this view publicly and privately. This interpretation suggests that without knowing, (or observing), an event, it's quantum state can be anything but will collapse upon measurement to some definite value. Other theories such as an infinite number of universes are "created" as a consequence of every event have also been proposed and supported by the scientific community. We choose to present a different view that is more in line with both Einstein and Schrödinger who came up with the theory in the first place.

We state simply that in order for an event to happen it requires an observer. We begin by first defining an event and then figuring out who the observer is.

Consider a very simple universe in which exists only a photon of light. The photon moves at the speed of light and is nothing more than a disturbance in the space-time continuum obeying the laws of physics as described by the Einstein Field Equations. Nothing happens. Nothing is going to happen. The photon just keeps travelling along.

Consider another very simple universe in which there is only a stone. The stone sits there and obeys all the laws of physics as described by the Einstein Field Equations. Nothing happens. Nothing is going to happen. The stone just sits there.

We define an event as something happens – an event occurs when something happens.

Consider now a universe with only a stone and a photon of light that is going to hit it. Both the stone and photon obey the laws of physics as described by the Einstein Field Equations and Schrödinger's Equation.

The photon hits and bounces off the stone. The stone recoils from the hit. Something happened. Who is the observer? In this universe exist only two things, the stone and the photon. We are not there. The photon is not actually a thing *per se*; it is merely a disturbance in the space-time continuum. However, the stone is a different matter. It causes tension in the field lines. For the sake of argument, we call this tension "curvature". It is not a perfectly formal definition of curvature, but we will use it to describe spacetime that is curved in space and time and also that there is tension along the field lines. It is rather complex when dealing with tensor fields, but let us call any spacetime that is not absolutely flat as being "curved" and the properties of such a region as it's "curvature." (This is an incredibly simplified view.)

Nevertheless, all we have is the stone and the photon. It may be possible to argue the photon is in some way the observer, but we reject the argument for now. However, the stone recoils from being hit. The photon just bounces off. The stone has reacted to changes in its local environment. The only candidate for an observer is the stone. The stone is the observer. Is the stone conscious or sentient? Does the stone perceive?

Perhaps perception, in very simple terms, can be intimated should something react to changes in its environment. If so, then in extremely basic terms, the stone perceives. Consciousness or sentience must wait until we ask the stone how it feels and it actually answers. We doubt that will happen. However, the stone is totally in tune with the changes of the universe. It reacts to everything around it instantly. Metaphysically it is pure consciousness. It is absolutely open and reactive to everything around it.

We say that anything that curves spacetime is a candidate for an observer. The universe observes itself. An electron observes its own interactions with electromagnetic and gravitational fields. Consciousness and sentience are not the domain of this discussion. Physics does not demand sentience or consciousness. However, it demands an observer. If you wish to know who is observing Schrödinger's Cat, ask the cat.

If there is no observer, there is nothing for the photon to bounce off of. If there is no observer, there can be no event. Nothing can happen. An observer bends spacetime. It is the curvature of spacetime, the bending of a path of light, a collision, acceleration, unbalanced forces, differences in potential and so on that cause things to happen. The observer is intrinsically involved in the event. Otherwise, it can't figure out if anything happened, whether it be a cat or something incredibly obedient, like a stone.

In summary, we find it somewhat incredible that a vast and overwhelmingly large number of intelligent and competent scientists would interpret the need for an observer to assume that we as human beings are the observer and nothing can happen in the universe without us. Religious thinkers believe this to be a scientific demand for the existence of a divine observer, or God. It isn't.

The universe exists. It always has. We are born into it. We are given the incredible gift of being able to observe it. We observe it and marvel. We pass on, and the universe continues to exist almost as though we had never been.

We have just begun to open the window to this marvellous cosmological panorama. Less than 100 years ago we were arguing over the existence of our very own galaxy – our back yard. We have just barely begin to look at the heavens which may well be infinite and eternal. There is a lot of colour out there. There are a lot of bizarre and new things we have never even imagined, let alone seen. It is absurd to think that in such a short time anyone on the face of the Earth has any idea of what we are seeing. A little humility might be in order.

Chapter 5

Jean-Paul Sartre

"All this talk by Hagel about existence is very fine,
but what does it have to do with Truth?", [Sartre (1956)]
-Jean-Paul Sartre

Sartre talks about Pierre who is not at the café and therefore must be. He has to be in order not to be at the café. And there is some talk about a table. Apparently tables are quite important for philosophy, [Sartre (1956)], [Russell (1902)], [Schopenhauer (1818)]. We do not present tables here nor someone who is not at a café in order to exist. Furthermore, we find Sartre a little deep, so we take the liberty to simplify.

Consider a stone. Ask yourself, "Does the stone exist because you perceive it to exist, or does your perception exist because you are looking at a stone?" Which causes which? Does the stone cause your perception of it to exist or does your perception magically cause the stone to appear and thereby exist?

Aristotle, often misquoted, states simply that if the stone ceases to exist, our perception of it ceases to exist along with it. He also states that if our perception of the stone ceases to exist, the stone continues to exist, [Aristotle (335 BC)]. In spite of all intervening arguments, there is the simple fact that the stone exists independently of our perception. Those who persist in believing a stone's existence is the result of their perception, or that reality is in any way the result of their perception, are mentally ill by definition. It is a fundamental fact for those who are sane that reality, or a stone, exists independently of perception.[1]

---

[1] A generation of youth who were reared in the 1960s have experimented with altering perception even to the point of death and perhaps beyond. All such experiments have resulted in the same conclusion: no matter how much perception is altered, reality does not change one iota.

If we therefore conclude that something as simple as a stone exists independently of our perception, then what about Truth?

To paraphrase the argument of Sartre, does a lie exist? Is there such a thing as a lie?

Has anyone at any time ever lied? Have you?

Is it possible to lie if there is no truth to lie about?

Alternatively, can there possibly exist a truth without anybody lying about it?

Given F = some falsehood and T = the truth being lied about, we have the following postulate:

$$\exists F \to \exists T. \tag{5.1}$$

However,
$$\exists T \to \neg \exists F \tag{5.2}$$
is more than possible.

Therefore, falsehood cannot exist without truth, but truth can exist without falsehood. Because of 5.2 we can pose:

$$\neg (\exists T \to \exists F) \tag{5.3}$$

So falsehood is not an anti-truth. It is possible that there may exist one truth, ie. a stone exists, and an infinite number of falsehoods concerning that truth. Namely:

- the stone does not exist

- the stone is not a stone

- the stone is only part of your imagination

- there are an infinite number of universes in which there are stones, this stone is not of this universe

There is a further interesting proof. Your perception is mostly in your brain which is located in your head. We could possibly take a stone and start hitting you repeatedly on your head to see if you could alter your perception of the stone in order to make it go away. There would be only one perception that would cause us to stop striking your head with the stone and that is that you perceive and agree that the stone actually exists independently of your perception. We call this technique, "proof by intimidation".

- the probability of the existence of the stone makes it to only *appear* to be here. If you wait long enough, it will disappear.

- the stone is only probably here; it's not really here.

- the local probability density is sufficient to cause instruments to measure the approximate spacial and temporal location of a region emitting quantum particles resembling a geophysical object, however the actual existence of a material quantum state is beyond the realm of science at this time

- by the authority invested in me by the supreme Magic Muffin – the repository of all knowledge and power – the existence of the stone is declared illegal and punishable by death

- ignore the stone, it will probably just go away and not bother us any more

- and so on ...

Nevertheless, even if there are a huge number of falsehoods, it takes but one truth to dissipate them all. Therefore, there is not a vast amount of truth and an equally vast amount of untruth or falsehood, such that when they meet they annihilate each other and there is nothing left. There is, for example, a vast ocean of falsehood and the simplest truth evaporates all of it immediately. Truth annihilates falsehood, but no falsehood can annihilate truth. Truth, therefore, exists. It exists in and of itself independently of either an anti-truth or our perception. A reality is truth. Truth is reality.

"Truth is that which exists independently of our perception."

- Cameron Rout

Continuing this line of thought, we note that it is impossible to get a wrong answer to a math exam if there is no right answer. However, the right answer can exist without there being any wrong answers.

Correctness can exist without error but error cannot exist without correctness.

Right can exist without wrong but wrong cannot exist without right.

What about Good and Evil? Is it that virtue can exist without vice but vice cannot exist without virtue?

19

To respond with a little tongue-in-cheek to Sartre's question: What does Truth have to do with existence? Very simply, existence is truth. Truth is what is. The closer you get to reality, the closer you get to the truth.

This all seems to make sense.

5.1 What Am I Trying to Do?

In this book, I am questioning everything. I am questioning our assumptions. As a matter of fact, I am going to assume the opposite of what others have assumed and see where it leads us. I am going to look at the very large and the very small. We have been led astray near the end of the twentieth century and the beginning of the 21st. And during this time, anyone who questions the assumptions of the academic elite is ostracized, ridiculed and blacklisted. We have ended up with a cosmology and epistemology that has been fabricated by madmen. I will tell you a story. Werner told it to me. It goes like this:

Werner's family started out in Germany and they had to leave because of the Second World War and the the the rise of fascism. Werner is Jewish and the genocide against the Jews forced Werner's parents to bundle up Werner as a baby and flee to South Africa where he grew up. After Werner had graduated from school, and according to what he told me, he became very precocious and wrote to none other than Erwin Schrödinger at Imperial College, Dublin, Ireland. That was where the great scientist was working at the time. Werner wrote that he wanted to work on the Unified Field Theory, in which both Erwin Schrödinger and Albert Einstein were so interested, and could he learn from Dr. Schrödinger. And much to Werner's surprise, he got a reply. But it wasn't from Schrödinger, it was from the dean of the college. Erwin Schrödinger had just retired and gone back to Germany. Nevertheless, Erwin had retired and was no longer in Ireland. However, the dean wrote, there was a position open to do research in general relativity and would Werner be interested? Werner told me, "Of course I was interested, and I would have been a complete idiot not to jump at the chance." So he packed up and left South Africa and got himself to Ireland.

In Ireland, Werner was given Erwin Schrödinger's old office. I guess you could say that Werner Israel was the first person to take the chair of Erwin Schrödinger, but Werner would be far too humble to be thought of as that. But it's kinda true. Nevertheless, when Werner moved into his office, the father of Quantum Mechanics had left all his stuff behind. He hadn't cleaned out his office and all his partially read papers he was working on were lying around. Werner sat in the chair of the office, probably after moving stuff off it, and just looked around. It must have been quite the experience for a young man so interested in the ways of science. He noticed a fairly large pile of papers on quantum gravity that Schrödinger had been going through with copious notes in the margins of the papers themselves. Of course the notes were in German, but Werner knew German and could follow the comments. He picked up the top paper and went through it, then another and another, exhausting the pile. And Werner told me, with a laugh, that all of the comments were pretty much the same. That is, all of the material on quantum gravity, according to Schrödinger, was the work of a bunch of raving lunatics. And after examining what I have personally found on quantum gravity, I am in complete agreement with Erwin Schrödinger. I imagine Werner is too.

Chapter 6

Schrödinger's Equation

Consider any particle with mass m and some electrostatic energy V which obeys the following:

$$i\hbar\frac{\partial}{\partial t}\psi = -\frac{\hbar^2}{2m}\nabla^2\psi + V\psi \tag{6.1}$$

or:

$$i\hbar\frac{\partial}{\partial t}\psi = -\frac{\hbar^2}{2m}\left\{\frac{\partial^2}{\partial x^2} + \frac{\partial^2}{\partial y^2} + \frac{\partial^2}{\partial z^2}\right\}\psi + V\psi \tag{6.2}$$

Let:
$$\psi = T(t)X(x)Y(y)Z(z) \tag{6.3}$$

where, $T(t)$ is a function of t only, $X(x)$ is a function of x only, $Y(y)$ is a function of y only and $Z(z)$ is a function of z only. Then:

$$i\hbar\frac{\partial}{\partial t}TXYZ = -\frac{\hbar^2}{2m}\left\{\frac{\partial^2}{\partial x^2}TXYZ + \frac{\partial^2}{\partial y^2}TXYZ + \frac{\partial^2}{\partial z^2}TXYZ\right\} \\ +VTXYZ \tag{6.4}$$

Under the condition that $\psi \neq 0$ we can divide through by $TXYZ$ to yield:

$$i\hbar\frac{T'}{T} = -\frac{\hbar^2}{2m}\frac{X''}{X} - \frac{\hbar^2}{2m}\frac{Y''}{Y} - \frac{\hbar^2}{2m}\frac{Z''}{Z} + V \tag{6.5}$$

We can see that each term is linearly independent. Since each term is being varied only by its independent variable and all terms are linearly independent from each other, and the constant term is also independent from the others, each term must equal a constant. Because we do not want this solution to blow up, we set the following:

$$i\hbar \frac{T'}{T} = -\alpha^2 \tag{6.6}$$

$$\frac{\hbar^2}{2m} \frac{X''}{X} = V - \beta^2 \tag{6.7}$$

$$\frac{\hbar^2}{2m} \frac{Y''}{Y} = -\gamma^2 \tag{6.8}$$

$$\frac{\hbar^2}{2m} \frac{Z''}{Z} = -\xi^2 \tag{6.9}$$

and the equation has been separated. We have placed the constant term with equation 6.7 since it has been chosen as the direction of travel of the particle. ψ is argued to be a measurement of the probability of the existence of the particle, which Schrödinger disagreed with rather strongly. We are going to show that a better interpretation of ψ is that it is a potential of some form, perhaps the potential of existence.

For the time-ordered term, we have,

$$T = e^{i \frac{\alpha^2}{\hbar} t} \tag{6.10}$$

which is an interesting equation. Consider:

$$e^{-k^2 \theta} \tag{6.11}$$

and k is a constant. If this were the case, then if k has any value, as θ increases, the expression will approach zero. So, if $\alpha \neq 0$ the time factor of ψ will act as a taper and kill the value of ψ. This would mean that the particle would evolve out of existence with the passage of time, somewhere along the order of $\hbar$. This would mean the particle would decay and all particles obeying this equation would decay and cease to exist. Obviously, this cannot be correct. We are re-deriving Schrödinger's original paper on this and we will end up with a fundamental principle:

existence is conserved.

24

We will show that the particle is stable and non-decaying.

Looking at the spacial characteristic function we have:

$$X = cos(\frac{\sqrt{2m(\beta^2 - V)}}{\hbar}x) \tag{6.12}$$

This is, in a way, similar to a term in a Fourier series. We set X in a π box[1] and consider a slight re-write as:

$$X = cos(\frac{2\pi\sqrt{2m(\beta^2 - V)}}{\hbar}\frac{x}{2\pi}) \tag{6.13}$$

This results in the boundary condition of the box in which $X = 1$ where;

$$x = \frac{2\pi\hbar}{\sqrt{2m(\beta^2 - V)}} \tag{6.14}$$

and

$$x\sqrt{2m(\beta^2 - V)} = h \tag{6.15}$$

since

$$\psi = 1 \tag{6.16}$$

we get

$$\mathbf{p}^2 = -\hbar^2\frac{\partial^2\psi}{\partial x^2} \tag{6.17}$$

at the boundary. We also have

$$-\frac{\hbar^2}{2m}\frac{\partial^2\psi}{\partial x^2} = \beta^2 - V \tag{6.18}$$

yielding

$$-\hbar^2\frac{\partial^2\psi}{\partial x^2} = 2m(\beta^2 - V) \tag{6.19}$$

Substitution yields:

$$x\mathbf{p} = h \tag{6.20}$$

at the boundary of the particle. However the "angle" in the π-box goes from 0 to 2π and therefore we have a measure of Δx. Because x varies between the boundaries we have a variable $\mathbf{p}$. We therefore have:

---

[1]From Fourier-series solutions to differential equations, we note the cos function is harmonic and repeats when the angle it is applied to extends from 0 to 2π. We consider two boundaries to the spacial region, one at angle 0 and the other at angle 2π. For example, consider $y = cos(u)$. The function y repeats and we consider boundaries of the region from $u = 0$ to $u = 2\pi$. Often I refer to this region as a π-box.

$$\Delta x \Delta \mathbf{p} = h \qquad (6.21)$$

We would like to mention here that the boundary happens to yield a potential of existence of one for the particle. Inside this boundary the potential of its existence is less than one. As a matter of fact, at the "centre" of the particle, the potential of existence is -1 and this appears at first to be absurd. In the derivation of a solution we had said $\psi \neq 0$. So we will deny the particle to exist inside the boundary and, for that matter, outside the boundary as well. For this particular solution to stand, the particle only exists where $\psi = 1$ and does not exist elsewhere. We are stating that the particle does not exist where $\psi < 1$. This is a different case than determining the position or time of the particle. In this case we are determining the existence of the particle itself. We are postulating that if ψ is less than one, then the particle isn't. We conclude ψ cannot be a measure of probability. It is a potential. When the potential is 1, the particle exists. From these calculations, the particle can only exist at it's boundary.

From outside the particle we have:

$$\Delta x \Delta \mathbf{p} > h \qquad (6.22)$$

Let us now return to the time ordered characteristic function, we have an exponential of $i\frac{\alpha^2}{\hbar}t$. Let us now reconsider α. We note the units of measure here. We see that $\hbar$ is in units of joules-sec. We see that t is in seconds and will cancel the time unit of $\hbar$ leaving joules in the denominator. Hence, since the exponential must be unitless, α^2 must be in units of joules. To continue the discussion allow α^2 to be some unknown form of energy. We will examine what this means as follows.

Let

$$E = \alpha^2 \qquad (6.23)$$

so the exponential of the time ordered factor becomes

$$\frac{iEt}{\hbar} \qquad (6.24)$$

and we look at the situation where $\psi = 1$. In other words, the particle definitely exists. We have seen that at the boundary of the box, from before, the spacial ordered factor is one. Therefore the time ordered factor is also one. This can only occur should the exponent of the time ordered factor be something like $2\pi i$. In which case we have:

26

$$\frac{iEt}{\hbar} = 2\pi i \tag{6.25}$$

rearranging

$$Et = 2\pi\hbar \tag{6.26}$$

$$Et = h \tag{6.27}$$

Here we have time going from 0 to some cyclic value yielding an exponent of $2\pi i$. We will then denote this as Δt and ΔE is the magnitude of fluctuation of energy. We now have:

$$\Delta E \Delta t = h \tag{6.28}$$

and observing from outside the particle in the time dimension, we write:

$$\Delta E \Delta t > h \tag{6.29}$$

This happens outside some time ordered "boundary" where/when the potential of the existence of the particle yields $\psi = 1$. Combining both time and spacial ordered factors we have the situation where

$$\Delta x \Delta \mathbf{p} \geq h \tag{6.30}$$

and

$$\Delta t \Delta E \geq h \tag{6.31}$$

Let us take a closer look at E.

The exponent of the time ordered factor is some phase angle that allows the particle to have a potential of existence equal to one at each cycle.

Let:

$$\frac{E}{\hbar}t = \theta \tag{6.32}$$

And we differentiate by t on each side to yield:

$$\frac{E}{\hbar} = \frac{d\theta}{dt} \tag{6.33}$$

or

$$\frac{E}{\hbar} = \omega \tag{6.34}$$

27

$$E = \hbar\omega \qquad (6.35)$$

and,

$$E = h\nu \qquad (6.36)$$

So, this energy, E, is not a form of energy coming from the mass of the particle or it's momentum of motion or even it's charge generating V. It appears to be an energy that is associated with the time ordered frequency of the particle's existence. This energy is not associated with mass or charge.

Let us examine α further.

$$\alpha = \frac{\sqrt{2\pi i \hbar}}{\sqrt{t}} \qquad (6.37)$$

and

$$\alpha = \frac{1}{\sqrt{2}} \left(\sqrt{\frac{h}{t}} \right) (1 + i) \qquad (6.38)$$

and it seems that with the presence of $\sqrt{2}$ there is some indication of spin involved.

Continuing, we see that we can also say:

$$\theta_n = n^2 2\pi i, \; n \in \mathbb{N} \qquad (6.39)$$

whenever ψ (or $T(t)$) $= 1$. So this exponential has been quantized by n^2. This can be compared to an orthogonal set of eigenfunctions yielding a complete solution of ψ.[2]

We now put forward a solution to the paradox of Schrödinger's cat.

---

[2]There are interesting consequences to the general solution of Schrödinger's equation. α is an eigenvalue in an eigenspace which can be used to find general solutions. Apparently $-\alpha^2$ is the energy of a photon. I am proposing that the magnitude of an infinite number of eigenvalues to the general solution of Schrödinger's equation yield the energy values of subatomic particles. The first order temporal eigenvalue yields the energy of a photon.

Chapter 7

Schrödinger's Cat

The cat is dead, [Sartre (1956)], [Python (1969)].[1]

---

[1]Schrödinger wrote: One can even set up quite ridiculous cases. A cat is penned up in a steel chamber, along with the following device (which must be secured against direct interference by the cat): in a Geiger counter there is a tiny bit of radioactive substance, so small, that perhaps in the course of the hour one of the atoms decays, but also, with equal probability, perhaps none; if it happens, the counter tube discharges and through a relay releases a hammer which shatters a small flask of hydrocyanic acid. If one has left this entire system to itself for an hour, one would say that the cat still lives if meanwhile no atom has decayed. The psi-function of the entire system would express this by having in it the living and dead cat (pardon the expression) mixed or smeared out in equal parts. It is typical of these cases that an indeterminacy originally restricted to the atomic domain becomes transformed into macroscopic indeterminacy, which can then be resolved by direct observation. That prevents us from so naively accepting as valid a "blurred model" for representing reality. In itself it would not embody anything unclear or contradictory. There is a difference between a shaky or out-of-focus photograph and a snapshot of clouds and fog banks.[1] The above text is a translation of two paragraphs from a much larger original article, which appeared in the German magazine Naturwissenschaften ("Natural Sciences") in 1935.[URL-10]

Chapter 8

The Field Equations

Consider an equation which partially comes from Minkowski and also quoted by Einstein, [Einstein (1914-17)]. This is the equation Dr. Werner Israel had given me many years ago.

$$G_{\mu\nu} = 8\pi T_{\mu\nu} - 2F_{\mu\alpha}F_\nu{}^\alpha + \frac{1}{2}g_{\mu\nu}F_{\alpha\beta}F^{\alpha\beta} \qquad (8.1)$$

From equation 8.1 we denote $T_{\mu\nu}$ as a material stress energy tensor and the Faraday tensor terms as a field stress energy tensor.

We can see that if there is no electric charge present we have the formula:

$$G_{\mu\nu} = 8\pi T_{\mu\nu} \qquad (8.2)$$

and we have the usual Einstein Field Equations for gravitational interactions.

Should there be no mass, but charge is present, we have:

$$G_{\mu\nu} = 2\left(-F_{\mu\alpha}F_\nu{}^\alpha + \frac{1}{4}g_{\mu\nu}F_{\alpha\beta}F^{\alpha\beta}\right). \qquad (8.3)$$

Equation 8.1 is the complete field equation resulting from the presence of both mass and charge in boundless space. Equation 8.2 is the gravitational field equation and equation 8.3 describes spacetime under Maxwell's Equations. Note that it was probably Minkowski who developed the tensor equation for Maxwell's electromagnetic theory and Einstein developed the tensor equation for gravity. For the sake of clarity, allow:

$$\Omega_{\mu\nu} = 2\left(-F_{\mu\alpha}F_\nu{}^\alpha + \frac{1}{4}g_{\mu\nu}F_{\alpha\beta}F^{\alpha\beta}\right) \qquad (8.4)$$

and

$$\Xi_{\mu\nu} = 8\pi T_{\mu\nu} \tag{8.5}$$

where $T_{\mu\nu}$ is a material stress-energy tensor[1]

So we have:

$$G_{\mu\nu} = \Xi_{\mu\nu} + \Omega_{\mu\nu} \tag{8.6}$$

and consider the following situation.[2]

In the case of a photon passing by the Sun, the mass of the Sun yields the material stress energy tensor already described. The fluctuations of the electromagnetic fields from the photon have to be derived from equation 8.3. The $\Omega_{\mu\nu}$ tensor is a microscopic view of the actions of a stationary object having electrostatic charge and magnetic properties. However, to describe the photon itself from these equations we need to take a macroscopic average of the stresses and energy generated by the $\Omega_{\mu\nu}$ tensor.

A photon is a region of rapidly fluctuating electric and magnetic fields moving at the speed of light. Overall, there is a rapidly changing (vibrating) stress-energy tensor within the region of the photon in which:

$$T_{\mu\nu} = \rho \mathbf{l}_\mu \mathbf{l}_\nu \tag{8.7}$$

where ρ is the energy density, or $h\nu$ per unit volume, of the photon, and l is the four-velocity of the particle of light known as a photon. In this way, it can be seen that the bundle of rapidly fluctuating electric and magnetic fields appears to behave on a macroscopic level as a particle having momentum and an incredibly small gravitational field. Therefore, if the appropriate differentiation is applied to the tensors describing a wave bundle moving at the speed of light, it's energy can be derived from equation 8.1 which must be equal to $h\nu$. In this way Plank's constant enters the field equations.

We know that $G_{\mu\nu}$ describes the curvature of a local region – in this case the local region of a photon. The $(0,0)$ component is the localized energy density. Momentum, pressure and shear stress densities are also contained in the Stress-Energy tensor (i.e. $T_{\mu\nu}$), which has been well known for nearly a hundred years. In this way, the local energy and momentum densities of a localized region of space-time undergoing rapid fluctuations of electric and magnetic fields and

---

[1] The proper nomenclature is that both are stress-energy tensors and are really both $T_{\mu\nu}$. However I am using one of the stress-energy tensors as a material stress-energy tensor, while the other is a field stress-energy tensor. I am relying on your generosity to allow me to use $T_{\mu\nu}$ in two different ways.

[2] I would like to call $\Xi_{\mu\nu}$ the Hawking Tensor and $\Omega_{\mu\nu}$ the Israel Tensor. That way the Unified Field Theory, without Quantum Mechanics, can simply be expressed as Einstein equals Hawking plus Israel. It has a nice historical ring to it.

moving at the speed of light can be calculated from well understood mathematical principles and procedures. Obviously, from equation 8.3, the energy and momentum of a photon can be obtained. The photon has momentum.

As the photon travels deeper into the gravitational well of the Sun, we see that the energy of the photon increases; it blue shifts. As the photon bypasses the Sun and climbs back out of the gravitational well, it red shifts. In the reference frame of the Sun, assuming the Sun's position is unaffected, the red shift equals the blue shift as the photon follows the geodesic described by the gravitational field of the Sun. However, there is one small problem with this approach. We are working in a theoretical non-moving reference frame – that of the Sun. We are assuming the Sun is not moving in our laboratory universe having only the Sun and a particular photon. If such were the case, the Sun being treated as an inertial reference frame, then there would be no resultant red shift of the photon by-passing the Sun. But such is not the case.

The photon has momentum. We must take this into consideration. We must move to an inertial reference frame utilizing the total momentum of the Sun and photon. The total momentum of the system must remain constant. In this reference frame, there is a slight change in the photon's momentum due to its change in direction, this change in momentum must be subtracted from the Sun's change in momentum so that the total momentum remains constant. This takes energy which comes from the photon, which red shifts to make up for the gain in kinetic energy of the Sun. As a result, the bypassing of the photon causes the Sun to very slightly move which in turn alters the region of the photon's local value of $T_{\mu\nu}$ and thereby, the value of $\rho l_\mu l_\nu$ for the photon. We see that the slope of the gravitational well is not the same when the photon exits the well of the Sun as when it enters it. This is because the Sun very slightly shifts toward the photon as it passes. The well is steeper coming out of it than when entering it.

General Relativity meets Quantum Mechanics.

Since the Faraday tensors describing the local spacetime of a photon affect $G_{\mu\nu}$, a photon therefore adds to the curvature of local space-time and therefore interacts gravitationally with local objects; however, extremely slightly.

We can figure out the red shift of the photon by treating the interaction as a collision using Plank's formula or we can figure it out with momentum considerations resulting from rather difficult and complex operations and differentiations on the photon's quickly moving locality and the interaction it has in changing the Sun's momentum. There are two ways to solve the problem and both should be equal.

Furthermore, we can also see the derivation of the very slight red shift of light by-passing the Sun has been approximated as linear. Over great distances and with very large masses, this effect becomes more pronounced and non-linear. There is a cosine factor involved which comes into play the more and more the light is "bent". At great distances this would not behave as a linear function. Eventually, extremely distant light would be very much red shifted. This is not meant to explain the red shifting of galaxies, but rather the low ambient temperature of the local region of space[3].

We next work out the red shift using the approach of a collision.

Determining the red shift using Maxwell's and Einstein's equations is left as an exercise for the reader.

Homework Assignment #1

- Derive a formula for Plank's constant from the equations of General Relativity.

- Determine the red shift factor of a photon bypassing the Sun.

- Determine the red shift factor of a photon bypassing a galaxy.

- Without using approximations, determine the distance a photon must travel in this universe, given local galactic density, so that its total angular deflection equals $90°$.

- According to your calculations, how far would a photon have to travel so that its redshift would be great enough to "cool" the photon from yellow light to $2.7°K$?

- According to your calculations, how fast would objects appear to be receding if a Doppler effect were assumed for the red shift? (Assume about 800 Mpc. as the photon starting distance.)

- integrate the total red shift due to the effects of an expanding universe and assume an initial temperature of $6,000,000,000°$ C for the origin of local photons and a time of 13.6 billion years. What is the local temperature of the universe? (Assume a free expansion with the present accepted value of the Hubble constant.)

---

[3] About $2.7°K$

Chapter 9

A Particle/Photon Gravitational Interaction

9.1 Discussion

If a photon having momentum $\frac{h\nu}{c}$ bounces off a sphere "at rest" having mass M and is deflected by a small angle θ, then the sphere would gain momentum, [Einstein (1905)]. This would mean that the sphere would, in an ideal situation, move very, very slowly. Since the sphere has gained kinetic energy, according to the law of conservation of energy, the photon would lose energy equivalent to the kinetic energy gained by the sphere. As a result, the photon must red shift.

9.2 Elastic Collision

Let us examine this interaction as a gravitational slingshot between a photon and the Sun. A gravitational slingshot can be approximated as an elastic collision and compared to the interaction of a collision between billiard balls on a frictionless pool table. If the Sun is considered an incompressible billiard ball and a photon is considered as an incompressible cue ball barely touching the Sun in a so-called "kiss shot", then we can calculate a possible change in frequency as follows:

Let θ be the angle of the photon coming off the "kiss" compared to travelling in a "straight" line as if missing the hit. If the Sun has mass M and recoils with velocity V, the conservation of momentum demands:

$$\frac{h\nu}{c} = MVsin(\frac{\theta}{2}) + \frac{h\nu'}{c}cos(\theta) \tag{9.1}$$

for the "x" direction and:

$$MVcos(\frac{\theta}{2}) = -\frac{h\nu'}{c}sin(\theta) \tag{9.2}$$

for the "y" direction.

Substituting for MV from equation 2 into equation 1, we have:

$$\frac{h\nu}{c} = \frac{h\nu'}{c}cos(\theta) - \frac{h\nu'}{c}\frac{sin(\theta)sin(\frac{\theta}{2})}{cos(\frac{\theta}{2})} \tag{9.3}$$

which reduces to:

$$\nu = \nu'(\frac{sin(\theta)sin(\frac{\theta}{2})}{cos(\frac{\theta}{2})} + cos(\theta)) \tag{9.4}$$

For very small θ:

$$\nu \simeq \nu'(\frac{\theta^2}{2} + 1) \tag{9.5}$$

Let $\nu - \nu' = \Delta\nu$. Then, subtracting ν' from both sides:

$$\Delta\nu \simeq \nu'\frac{\theta^2}{2} \tag{9.6}$$

Since $\nu' \simeq \nu$ we have:

$$\Delta\nu \simeq \nu\frac{\theta^2}{2} \tag{9.7}$$

9.3 Calculations from General Relativity

The first approximation of a solution to a solar/photon interaction was done by Einstein, [Einstein (1905)] in which the following equation was derived:

$$\theta_{rad} = \frac{4M}{R} \tag{9.8}$$

in which θ_{rad} is the angle coming off the interaction in radians, M is the mass of the Sun in meters, (Schwartzchild radius). The mass of the Sun is about $M = 1475$ meters. R is the distance from the centre of the Sun to the point of perihelion of the hyperbolic orbit of the photon in the path around the Sun following the geodesic, also in meters. To convert R to the angular separation from the star to the Sun in radians, divide R by an astronomical unit.

The formula derived previously is:

$$\Delta\nu = \frac{\theta^2_{deg}}{2}\nu \tag{9.9}$$

where θ_{deg} is the angle coming off the interaction in degrees, $\Delta\nu$ is the change in frequency and ν is the frequency of the signal.

Combine both formulae to yield:

$$\Delta\nu = \frac{M^2\pi^2}{2025R^2}\nu \tag{9.10}$$

or:

$$\Delta E = \frac{M^2\pi^2}{2025R^2}E \tag{9.11}$$

or:

$$\Delta p = \frac{M^2\pi^2}{2025R^2}p \tag{9.12}$$

By-passing the Sun, a photon is red-shifted by a factor of about 10^{-7}.

Chapter 10

String Theory

These days, it seems, everyone is investigating String Theory as a possible path to a Unified Theory. It shows a lot of promise. Since I thoroughly encourage off-the-wall, out-of-left-field, and completely bizarre approaches to theories of physics, I shall take a little walk into this strange world. Please note that this work is complete conjecture and probably has no chance of experimental verification. Now, that's my kind of physics!

We have looked at Schrödinger's Equation, which is a diffusion equation with a linear term added on the end. Diffusion is an interesting mathematical phenomena. Mathematically, effects occur instantly throughout the media into which diffusion is penetrating. That means things happen faster than the speed of light; they happen instantly. This is just a property of the diffusion equation.

Consider three classes of differential equations. The diffusion equation, of which Schrödinger is a part, the harmonic equation and the biharmonic equation.

Setting these out:

$$\text{The diffusion equation:} \quad \frac{\partial \psi}{\partial t} = \nabla^2 \psi$$

$$\text{The harmonic equation:} \quad \frac{\partial^2 \psi}{\partial t^2} = \nabla^2 \psi \quad (10.1)$$

$$\text{The biharmonic equation:} \quad 0 = \nabla^4 \psi$$

They look kind of pretty, don't they? That basically covers the field of applied mathematics as we know it. There are many variations of these three formulae which make the world of science more interesting and challenging.

39

But these three cover most of it. By the way, ψ is almost always defined as some unknown potential We're mathematicians; we don't care about the units.

These three equations cover a great deal within the field of applied mathematics. There are, of course, many variations.

The second formula above, the harmonic equation, has the property that alterations propagate at a particular speed. That the speed of propagation within the media described by the differential equation has some definite finite value such as the speed of light. However, space-time is a tensor field and the harmonic equation can only describe a vector field. This has possibilities for electromagnetism but not for gravity.

The third formula above is the biharmonic equation and describes the world of elasticity. It is used in geophysics to describe movements of plate tectonics. It involves stress tensors of an elastic media.

Consider the Heisenberg shell previously derived[1]. Consider a potential ψ within a shell bounded by $r = \frac{h}{2mc}$ and $t = \frac{h}{mc^2}$. If we use the Schrödinger Equation as follows:

$$\left[i\frac{\partial}{\partial t} \right] \psi = \left[-\frac{\hbar^2 \nabla^2}{2m} + V \right] \psi \tag{10.2}$$

as a "Schrödinger operator". In order to find a measure of acceleration or force, or second time ordered differential, we re-apply the operator within the shell to obtain:

$$-\frac{\partial^2 \psi}{\partial t^2} = \frac{\hbar^4}{4m^2} \nabla^4 \psi - \frac{\hbar^2 V}{m} \nabla^2 \psi + V^2 \psi \tag{10.3}$$

And I have skipped a lot of steps. Sorry. Let me explain please.

Consider Schrödingers Equation written this way:

$$\left[i\hbar \frac{\partial}{\partial t} \right] \psi = \left[-\frac{\hbar^2}{2m} \nabla^2 + V \right] \psi \tag{10.4}$$

then cancel ψ on both sides to get:

$$\left[i\hbar \frac{\partial}{\partial t} \right] = \left[-\frac{\hbar^2}{2m} \nabla^2 + V \right] \tag{10.5}$$

---

[1] Where Bruce? Which previous chapter? What the heck is a Heisenberg shell??? Ok, consider a four-dimensional spherical shell in which $\Delta x = 2\Delta r$ and $r = \sqrt{x^2 + y^2 + z^2}$ and notice that I have used x to represent two different things. I'm not supposed to do that. So let's use $\Delta d = 2\Delta r$ instead to take in the spacial coordinates and we got a sphere radius r and diameter d and we are confined in time Δt, ok? The conditions of $\Delta t \Delta E = h$ and $\Delta d \Delta p = h$ form what I am calling a Heisenberg sphere; but I call it a Heisenberg shell in this case.

then square both sides to get:

$$\left[-\frac{\partial^2}{\partial t^2} \right] = \left[\frac{\hbar^4}{4m^2} \nabla^4 - \frac{\hbar^2 V}{m} \nabla^2 + V^2 \right] \tag{10.6}$$

then multiply both sides by ψ to get equation (10.3).

We look again at the Einstein Field Equations:

$$G_{\mu\nu} = 8\pi T_{\mu\nu} - 2F_{\mu\alpha}F_\nu{}^\alpha + \frac{1}{2}g_{\mu\nu}F_{\alpha\beta}F^{\alpha\beta} \tag{10.7}$$

and from the Principle of Equivalence[2] there exists some mapping whereby:

$$G_{\mu\nu} \Longrightarrow -\frac{\partial^2 \psi}{\partial t^2} \tag{10.8}$$

and since gravity is a fourth-ordered differential, a mapping exists whereby

$$8\pi T_{\mu\nu} \Longrightarrow \frac{\hbar^4}{4m^2} \nabla^4 \psi \tag{10.9}$$

and from the second ordered differentials of Maxwell's Equations we can find a mapping whereby

$$-2F_{\mu\alpha}F_\nu{}^\alpha + \frac{1}{2}g_{\mu\nu}F_{\alpha\beta}F^{\alpha\beta} \Longrightarrow -\frac{\hbar^2 V}{m}\nabla^2\psi + V^2 m \tag{10.10}$$

for some ψ.

Wrapping this all together and applying Schrodinger's equation to itself and then substitute into the above mappings to combine Maxwell's Equation, with Gravity and Quantum Mechanics, in spherical coordinates we venture to propose the following equation:

$\frac{\partial^2}{\partial t^2}\psi(t,r,\theta,\phi) = \frac{\hbar^4}{4m^2}$ (2 r sin(θ) (- 2 (2 r sin(θ) $\frac{\partial}{\partial r}\psi(t,r,\theta,\phi)$ + r^2

sin (θ) $\frac{\partial^2}{\partial r^2}\psi(t,r,\theta,\phi)$ + cos (θ) $\frac{\partial}{\partial\theta}\psi(t,r,\theta,\phi)$ + sin (θ) $\frac{\partial^2}{\partial\theta^2}\psi(t,r,\theta,\phi)$ +

$\frac{\frac{\partial^2}{\partial\phi^2}\psi(t,r,\theta,\phi)}{\sin(\theta)}$) $\frac{r^{-3}}{\sin(\theta))}$ + (2 sin (θ) $\frac{\partial}{\partial r}\psi(t,r,\theta,\phi)$ + 4 r sin (θ) $\frac{\partial^2}{\partial r^2}\psi(t,r,\theta,\phi)$

+ r^2 sin (θ) $\frac{\partial^3}{\partial r^3}\psi(t,r,\theta,\phi)$ + cos (θ) $\frac{\partial^2}{\partial r\partial\theta}\psi(t,r,\theta,\phi)$

+ sin (θ) $\frac{\partial^3}{\partial\theta\partial r\partial\theta}\psi(t,r,\theta,\phi)$

+ $\frac{\frac{\partial^3}{\partial\phi^2\partial r}\psi(t,r,\theta,\phi)}{\sin(\theta)}$) r^{-2} (sin (θ))$^{-1}$) + r^2 sin (θ) (6 (2 r sin (θ) $\frac{\partial}{\partial r}\psi(t,r,\theta,\phi)$

+ r^2 sin (θ) $\frac{\partial^2}{\partial r^2}\psi(t,r,\theta,\phi)$ + cos (θ) $\frac{\partial}{\partial\theta}\psi(t,r,\theta,\phi)$ + sin (θ) $\frac{\partial^2}{\partial\theta^2}\psi(t,r,\theta,\phi)$

---

[2]Accelerating reference frames and gravitational fields are equivalent.

$+ \dfrac{\frac{\partial^2}{\partial\phi^2}\psi(t,r,\theta,\phi)}{\sin(\theta)}$) r^{-4} $(\sin(\theta))^{-1}$ - 4 (2 $\sin(\theta)$ $\frac{\partial}{\partial r}\psi(t,r,\theta,\phi)$ +4 $r\sin(\theta)$

$\frac{\partial^2}{\partial r^2}\psi(t,r,\theta,\phi) + r^2\sin(\theta)\frac{\partial^3}{\partial r^3}\psi(t,r,\theta,\phi) + \cos(\theta)\frac{\partial^2}{\partial r\partial\theta}\psi(t,r,\theta,\phi)$

$+ \sin(\theta)\frac{\partial^3}{\partial\theta\partial r\partial\theta}\psi(t,r,\theta,\phi) + \dfrac{\frac{\partial^3}{\partial\phi^2\partial r}\psi(t,r,\theta,\phi)}{\sin(\theta)}$) $r^{-3}\sin(\theta)^{-1}$

$+ 6\sin(\theta)\frac{\partial^2}{\partial r^2}\psi(t,r,\theta,\phi) + 6r\sin(\theta)\frac{\partial^3}{\partial r^3}\psi(t,r,\theta,\phi)$

$+ r^2\sin(\theta)\frac{\partial^4}{\partial r^4}\psi(t,r,\theta,\phi) + \cos(\theta)\frac{\partial^3}{\partial r^2\partial\theta}\psi(t,r,\theta,\phi)$

$+ \sin(\theta)\frac{\partial^4}{\partial\theta\partial r^2\partial\theta}\psi(t,r,\theta,\phi) + \dfrac{\frac{\partial^4}{\partial r\partial\phi^2\partial r}\psi(t,r,\theta,\phi)}{\sin(\theta)}$) r^{-2} $(\sin(\theta))^{-1}$) $+ \cos(\theta)$ (-

(2 $r\sin(\theta)$ $\frac{\partial}{\partial r}\psi(t,r,\theta,\phi) + r^2\sin(\theta)\frac{\partial^2}{\partial r^2}\psi(t,r,\theta,\phi) + \cos(\theta)\frac{\partial}{\partial\theta}\psi(t,r,\theta,\phi)$

$+ \sin(\theta)\frac{\partial^2}{\partial\theta^2}\psi(t,r,\theta,\phi) + \dfrac{\frac{\partial^2}{\partial\phi^2}\psi(t,r,\theta,\phi)}{\sin(\theta)}$) $\cos(\theta)$ r^{-2} $(\sin(\theta))^{-2}$ $+ 2r\cos(\theta)$

$\frac{\partial}{\partial r}\psi(t,r,\theta,\phi) + 2r\sin(\theta)\frac{\partial^2}{\partial r\partial\theta}\psi(t,r,\theta,\phi) + r^2\cos(\theta)\frac{\partial^2}{\partial r^2}\psi(t,r,\theta,\phi) + r^2$

$\sin(\theta)\frac{\partial^3}{\partial r^2\partial\theta}\psi(t,r,\theta,\phi)$ - $\sin(\theta)\frac{\partial}{\partial\theta}\psi(t,r,\theta,\phi) + 2\cos(\theta)\frac{\partial^2}{\partial\theta^2}\psi(t,r,\theta,\phi)$ +

$\sin(\theta)\frac{\partial^3}{\partial\theta^3}\psi(t,r,\theta,\phi)$ - $\dfrac{\left(\frac{\partial^2}{\partial\phi^2}\psi(t,r,\theta,\phi)\right)\cos(\theta)}{(\sin(\theta))^2} + \dfrac{\frac{\partial^3}{\partial\phi^2\partial\theta}\psi(t,r,\theta,\phi)}{\sin(\theta)}$) r^{-2} ($\sin(\theta)$

)$^{-1}$) $+ \sin(\theta)$ (2 (2 $r\sin(\theta)$ $\frac{\partial}{\partial r}\psi(t,r,\theta,\phi) + r^2\sin(\theta)\frac{\partial^2}{\partial r^2}\psi(t,r,\theta,\phi)$ +

$\cos(\theta)\frac{\partial}{\partial\theta}\psi(t,r,\theta,\phi) + \sin(\theta)\frac{\partial^2}{\partial\theta^2}\psi(t,r,\theta,\phi) + \dfrac{\frac{\partial^2}{\partial\phi^2}\psi(t,r,\theta,\phi)}{\sin(\theta)}$) $(\cos(\theta))^2$ r^{-2}

($\sin(\theta)$)$^{-3}$ -2 (2 $r\cos(\theta)$ $\frac{\partial}{\partial r}\psi(t,r,\theta,\phi) + 2r\sin(\theta)\frac{\partial^2}{\partial r\partial\theta}\psi(t,r,\theta,\phi) + r^2$

$\cos(\theta)\frac{\partial^2}{\partial r^2}\psi(t,r,\theta,\phi) + r^2\sin(\theta)\frac{\partial^3}{\partial r^2\partial\theta}\psi(t,r,\theta,\phi)$ - $\sin(\theta)\frac{\partial}{\partial\theta}\psi(t,r,\theta,\phi)$

$+ 2\cos(\theta)\frac{\partial^2}{\partial\theta^2}\psi(t,r,\theta,\phi) + \sin(\theta)\frac{\partial^3}{\partial\theta^3}\psi(t,r,\theta,\phi)$

$- \dfrac{\left(\frac{\partial^2}{\partial\phi^2}\psi(t,r,\theta,\phi)\right)\cos(\theta)}{(\sin(\theta))^2} + \dfrac{\frac{\partial^3}{\partial\phi^2\partial\theta}\psi(t,r,\theta,\phi)}{\sin(\theta)}$) $\cos(\theta)$ r^{-2} ($\sin(\theta)$)$^{-2}$ $+$ (2 r

$sin(\theta)$

$\frac{\partial}{\partial r}\psi(t,r,\theta,\phi) + r^2\sin(\theta)\frac{\partial^2}{\partial r^2}\psi(t,r,\theta,\phi) + \cos(\theta)\frac{\partial}{\partial\theta}\psi(t,r,\theta,\phi) + \sin(\theta)$

$\frac{\partial^2}{\partial\theta^2}\psi(t,r,\theta,\phi) + \dfrac{\frac{\partial^2}{\partial\phi^2}\psi(t,r,\theta,\phi)}{\sin(\theta)}$) r^{-2} $(\sin(\theta))^{-1}$ $+$ (- 2 $r\sin(\theta)$ $\frac{\partial}{\partial r}\psi(t,r,\theta,\phi)$

$+ 4r\cos(\theta)\frac{\partial^2}{\partial r\partial\theta}\psi(t,r,\theta,\phi) + 2r\sin(\theta)\frac{\partial^3}{\partial\theta\partial r\partial\theta}\psi(t,r,\theta,\phi)$ - $r^2\sin(\theta)$

$\frac{\partial^2}{\partial r^2}\psi(t,r,\theta,\phi) + 2r^2\cos(\theta)\frac{\partial^3}{\partial r^2\partial\theta}\psi(t,r,\theta,\phi)$ +

$r^2\sin(\theta)\frac{\partial^4}{\partial\theta\partial r^2\partial\theta}\psi(t,r,\theta,\phi)$ - $\cos(\theta)\frac{\partial}{\partial\theta}\psi(t,r,\theta,\phi)$

- 3 $\sin(\theta)\frac{\partial^2}{\partial\theta^2}\psi(t,r,\theta,\phi) + 3\cos(\theta)\frac{\partial^3}{\partial\theta^3}\psi(t,r,\theta,\phi) + \sin(\theta)\frac{\partial^4}{\partial\theta^4}\psi(t,r,\theta,\phi)$

$+2\dfrac{\frac{\partial^2}{\partial\phi^2}\psi(t,r,\theta,\phi)\cos(\theta)^2}{\sin\theta)^3}$ - 2 $\dfrac{\cos(\theta)\frac{\partial^3}{\partial\phi^2\partial\theta}\psi(t,r,\theta,\phi)}{(\sin(\theta))^2} + \dfrac{\frac{\partial^2}{\partial\phi^2}\psi(t,r,\theta,\phi)}{\sin(\theta)}$

$+ \dfrac{\frac{\partial^4}{\partial\theta\partial\phi^2\partial\theta}\psi(t,r,\theta,\phi)}{\sin(\theta)}$) r^{-2} $(\sin(\theta))^{-1}$) $+$ (2 $\sin(\theta)$ $\frac{\partial^3}{\partial\phi^2\partial r}\psi(t,r,\theta,\phi) + r^2$

$\sin(\theta)\frac{\partial^4}{\partial r\partial\phi^2\partial r}\psi(t,r,\theta,\phi) + \cos(\theta)\frac{\partial^3}{\partial\phi^2\partial\theta}\psi(t,r,\theta,\phi)$ +

$\sin(\theta)\frac{\partial^4}{\partial\theta\partial\phi^2\partial\theta}\psi(t,r,\theta,\phi) + \dfrac{\frac{\partial^4}{\partial\phi^4}\psi(t,r,\theta,\phi)}{\sin(\theta)}$) r^{-2} $(\sin(\theta))^{-2}$) r^{-2} $(\sin(\theta))^{-1}$

$- \dfrac{\hbar^2 V}{m}$ (2 r $\sin(\theta)$ $\frac{\partial}{\partial r}\psi(t,r,\theta,\phi) + r^2\sin(\theta)\frac{\partial^2}{\partial r^2}\psi(t,r,\theta,\phi) + \cos(\theta)$

$$\frac{\partial}{\partial\theta}\psi\left(t,r,\theta,\phi\right)+\sin\left(\theta\right)\frac{\partial^2}{\partial\theta^2}\psi\left(t,r,\theta,\phi\right)+\frac{\frac{\partial^2}{\partial\phi^2}\psi(t,r,\theta,\phi)}{\sin(\theta)}\,\Big)\,r^{-2}\,(\sin\left(\theta\right))^{-1}$$
$$+\,V^2\,\psi\left(t,r,\theta,\phi\right)$$

From the general solution to the above equation we can apply boundary conditions: $r=\frac{h}{2mc}$ and initial condition $t=\frac{h}{mc^2}$ to show that the resultant Bessel functions and their zeros along with Legendre polynomials lead to zeta functions appropriate to develop string theory.

Homework Assignment #2

1. Find the general solution to the above equation.

2. Apply boundary conditions: $r=\frac{h}{2mc}$ and initial condition $t=\frac{h}{mc^2}$ to your solution.

3. Show the resultant Bessel functions and their zeros along with Legendre polynomials lead to zeta functions and develop string theory. Show your work.

Chapter 11

Galaxies and the Unified Field Theory

We have travelled some way to get here. I have presented a Unified Field Theory. My son tells me that there can only be one Unified Field Theory, that it exists and that it is unique. Therefore, let us say, I have found the all important Unified Field Theory. And apart from the exotic, fantastic and bizarre theories previously postulated as overall explanations of physical reality, we have neatly tied up a rather straightforward, mundane and boring theory that elegantly combines the three great closed theories of the past into a very simple package; that they are all part of one, overall, complete and closed theory of physics.

These three theories are Gravity, Electromagnetism and Quantum Mechanics. You may think that only Gravity and Electromagnetism are closed theories and that Quantum Mechanics is still open, but you would be in error. Quantum Mechanics became a closed theory the moment the general time-dependant Schrödinger Equation with Heisenberg boundary conditions was solved. What does this mean?

First, please allow me to state that I am absolutely stunned and gob-smacked that a reader would trudge and grapple his or her way through endless chapters of obtuse physical theory and mathematical derivations and not have a clue why they were doing so. And, to let you in on a little secret, I have no idea why I am writing this either until I actually get to writing it down. I have taken a break in writing this treatise for about two years until I could figure out what to write next. And I believe I have come to the point where we must pause, look around and answer the age-old question: "Where the heck are we?" We know that we are going from darkness into light. And that is always a worthwhile journey. For example, when I was locked up in my room for years trying to trisect an angle, I thought it was possible to do even though I was told it wasn't. Then, in second year university, the proof that one cannot trisect an angle was presented. I think it was supposed to be by Minkowski or perhaps by Riemann. And, believe me, Minkowski and Riemann were geniuses. I have now found out that Archimedes trisected angles by using an Archimedes spiral millennia ago. So the guy beat me to it. I always thought the proof that it was impossible was convoluted and, in a word, crap. But now I know for sure. Minkowski and Riemann never postulated such a "proof". It is all a lie. Archimedes was the greatest mathematician who ever lived. Nevertheless, it was by attempting the impossible that I have gained an appreciation for genius. I don't exactly know where we are going with this, but we are going somewhere. And somewhere else is better than where we are now.

If I may explain, Schrödinger's Equation is a description of the potential of space and time within certain boundary limits. Within these limits, we can describe the physical state and properties of this region as though we were dealing with a particle. These particles are considered as within and of themselves, independently of other particles, and without any causation of events before and after. They just be as they are. And their description is determined by the eigenvalues within the solution of the equation known as Schröginer's Equation. When α^2 has the value of $2\pi i/h$, or $n = 1$, and we are dealing with the time oriented solution, we have a photon. In other words, $E = h\nu$ where ν is a collection of frequencies centred around a signature, or central frequency, that is determined by its energy. Both its energy and momentum are determined by the boundary conditions predicted by Heisenberg.

However, outside of this boundary, we utilize an entirely different set of differential equations known as the Einstein Field Equations which also include Maxwell's Equations. It is still all the same time and space, but time and space behave according to the the measure involved. Or, to be more correct, whether or not a measure actually exists. If we are measuring time and space within the Heisenberg boundary conditions, we are dealing with Schrödinger and Quantum Mechanics, which is not a Banach space; a measure does not exist. Outside of these boundary conditions, we deal with Einstein, Maxwell, Gravity and Electromagnetism which is a Banach space and measures exist. Where there is no particle, there is time and space and a measure of ψ; where we are dealing with a particle, there is a different measure of time and space and ψ. If we are dealing with particles that are described by α_n where:

$$i\hbar\frac{dT(t)}{dt} = -\alpha_n^2 T(t),$$

46

where $T(t)$ is the time ordered characteristic function of ψ, we have a "mass-less" particle that moves at the speed of light. If this wave bundle, which is the collection of all of the frequencies described by α_n, are treated as a particle having energy $h\nu$, then we can treat this particle within the larger bounds of space-time as:

$$T_{\mu\nu} = \rho l_\mu l_\nu$$

or,

$$G_{\mu\nu} = 8\pi\rho l_\mu l_\nu$$

where l is the four-velocity of the bundle. In our case, c. Here ρ is the energy density found by applying the Lagrangian to the Faraday tensor arrangement previously presented. Or,

$$G_{\mu\nu} = -2F_{\mu\alpha}F_\nu{}^\alpha + \frac{1}{2}g_{\mu\nu}F_{\alpha\beta}F^{\alpha\beta}$$

where α and β are not eigenvalues, but dummy indexes. Sorry for using the same variable to mean two different things, like α, but I am trying to follow convention.

The value of ψ is a measure of the potential of existence. This existence comes into being where $\psi = 1$. This is at the boundary of the particle. Space and time can be bent or curved by various forces. The measure of this curvature is the value of $G_{\mu\nu}$. We can see the effects of this curvature as a measure of acceleration, or as $\partial^2\psi/\partial t^2$. As spacetime is bent more and more by the application of greater forces, the value of ψ increases. And it seems you can only bend spacetime so far. If you bend it too much it "crimps" and becomes a particle, or a space-time knot, if you like, having the Heisenberg boundary conditions and an interior described by Schrödinger. At the boundary, spacetime "crystallizes" into mass and charge described by the boundary conditions which I have previously presented.

We must now find some physical phenomenon or some evidence somewhere that proves this true. Even though I have described the red shifting of light travelling through a gravitational field, that is a difficult measurement. There is another direct piece of evidence that can be easily seen by everyone. And that evidence is presented by galaxies. Spiral galaxies, and galaxies in general, have presented modern science with the greatest enigma in history. And it is through the resolution of this enigma that we can firmly establish the validity of the Unified Field Theorem. Yummy.

The enigma is that the scientific community at large believes there was a beginning to the universe about 13.6 billion years ago and that the universe has been expanding ever since. This is balderdash. Lately it has been postulated that the universe is not only expanding but that it is accelerating outwardly. Further, there has also been postulated that there exists an exotic substance and ethereal fantasy known together as dark matter and dark energy. This results in a cosmology that is both bizarre and completely mad. There is no dark matter, the universe is not expanding, there was no big bang. The universe is infinite: eternal in the past, eternal in the future. We can measure the distances to galaxies and show there is no analytical relationship to galactic red shift which proves that the universe is not expanding. And we can demonstrate that the universe can break the second law of thermodynamics to prove the universe is eternal. I shall show that the measure of the distance to galaxies proves the universe is not expanding. Actually, the universe is not accelerating outwardly from our galaxy; we are accelerating towards the centre of the galaxy away from the rest of the universe as a result of our circular motion around the galaxy. The demonstration of the ability of the universe to reverse the flow of entropy has been done through the unique and iridescent genius of Dr. Werner Israel.

Werner and I are on opposite ends of the spectrum. Werner is as conservative as you could imagine. An absolute gentleman, a kindly and pure soul. I, on the other hand, am some what of a social bull in a china shop. I try to spare Werner too much exposure to the ethereal realm of my mathematical indulgences. Actually, I have only met with him twice in recent years, and then I got very excited about some stuff, like finding out there is no dark matter, and it was at the same time Werner lost his son. It must have been devastating. I remember the time I went to Victoria a couple of years ago and met with Werner in his office. Things started out rather pleasantly and soon deteriorated into an incredibly polite argument over billiard balls. It was to demonstrate that light had to red shift as it passed through a gravitational field. I felt I had held my own. During the conversation, Werner had proposed a theory based on the expansion of the universe. I interrupted with the question, "What happens to your theory if the universe is not expanding?" To date there has been no answer to that question.

However, if the universe is not expanding and has been around forever, what happens to the second law of thermodynamics? If entropy always increases, then the universe must run down. It must eventually run out of fuel. How do we get around this?

The second time I saw Werner in recent times, I had postulated the existence of an upper limit to the mass of a black hole. That, perhaps, at reaching the mass of a galaxy, a black hole explodes.

Werner, patiently listening to me sitting in his office, was taken aback at hearing this postulate.

"And what could possibly cause a black hole to explode?" he asked quite incredulously.

"I was thinking," I said, "that it somehow hits absolute zero. That it bounced off of a thermodynamic barrier and exploded. Perhaps entropy reversed somehow and the black hole explodes to form another galaxy."

"Oh," replied Werner kindly, "Black holes are already very cold. They are around 10^{-20} degrees Absolute."

"Yes," I ventured, "but it is not absolute zero. It is not exactly zero. What is the temperature of a black hole?"

"One over M," he replied.

"Ah," I said, "it is a dead end. You would need an infinite mass to hit absolute zero."

Werner sat and thought. I waited. (Very quietly).

"It would be possible," he finally said, "for a black hole to actually hit absolute zero if it was spinning or if it had charge."

Hey, charge I don't care about, but everything has got to spin. But I kept quiet.

Werner concentrated.

"It would hit absolute zero if its angular momentum was equal to its mass squared," he concluded, fairly satisfied. I reeled from the on-coming tidal wave of mathematics that I was going to have to go through to follow that thought. However, Werner looked at me and smiled, "Of course that is in these silly relativistic unit-less units."

Of course.

I began to think and mutter about spinning black holes, but Werner was lost in thought.

"It would have to be spinning very fast," he said, still thinking. I held my breath. "It would have to be spinning so fast," he added, "that the surface of the black hole at its equator would be moving at the speed of light."

Bingo.

Let us now look at the structure of galaxies themselves and then work our way into their centre.

11.1 Quaesars and Gallactic Formation

Here we are going out on a limb. We have presented a postulate that the universe may not be expanding after all and shown that there is a connection between quantum mechanics and general relativity. This can be observed in the red shifting of light as it passes through a gravitational field. Let us look at very exotic cosmic entities which are called quasars – quasi-stellar radio objects. They are quasi-stellar because they are very small. They are radio objects since they are detected with radio telescopes and are not visual objects as of this writing. Although recently a picture has been taken of one. It looks like a very bright star. Some cosmologists believe they may be involved in the formation of galaxies.

Some facts about quasars include:

- they are either coming at us or going away from us at speeds close to the speed of light

- they have a period of luminosity of about a month and a half

- they can't be any bigger than one and a half light months across

- the Schwartzchild radius of our galaxy is about the distance light can travel in a month and a half

- they give off 40 times more light, at least, than an entire galaxy

- they may have absolutely nothing to do with galaxies (Actually, latest evidence shows this is not true. They may be galaxies being "reborn". But the rest of these points are still valid. Honest.)

Another major problem being worked on within the fields of quantum mechanics and cosmology is that there is a lack of anti-particles in our universe.

11.1.1 Sines, Cosines and Antimatter

Consider the following: In the spacial dimensions to the solution to the Schrödinger Equation, there exists a cosine function with an infinite number of eigenvalues. Looking at the cosine function, if we set the boundaries at $x = 0$ and 2π, then $\psi = 1$ at the boundaries. However, in the middle, $x = \pi$, and has the value of -1. Does this have a physical meaning in terms of the potential of existence? What could the meaning be of $\psi < 0$? Suppose such regions produce anti-matter. If we have a sphere and we compare the volume inside half the radius

50

with the volume on the outer half, we see that 1/8 of the volume is contained by the negative part and 7/8 is in the positive section of the sphere. If the negative potential ends up, somehow, as anti-matter and the positive part ends up as matter, there would be the annihilation of 1/4 of the volume of the sphere through matter-antimatter reaction and 3/4 of the volume of the sphere would end up surviving as matter.[1] Continuing on, if we had something like a black hole having the mass of a galaxy and it explodes for some unknown reason, like a super-exotic cosmic sub-atomic particle, we could possibly have the energy output of the annihilation in the order of a quarter of the mass of a galaxy. If true, that's gotta light up the night sky somewhere.

11.1.2 Academic Handwaving

Let's also look a little more at galaxy formation. If a computer model attempts to construct the formation of a galaxy, interesting things happen. In particular, it is difficult to construct the evolution of the galactic arms. If the arms are programmed to evolve or grow outwardly from a central core, they are fairly easy to be made stable through various programming methods. For a while it was impossible to get the galactic arms to be stable from a model based on the collapse of a galactic dust cloud. Eventually the problem was solved so that a dust cloud collapse can form stable galactic arms, but it took a while. It's a lot easier to do starting with a central blob of stars and letting the arms grow outwardly. Another interesting thing about galaxies is the halo. The halo, or region II of the galaxy, is populated with globular clusters. A globular cluster is a truly beautiful glob of millions of stars all buzzing about a central region like bees about their hive. It has no inter-stellar dust or gas. Globular clusters are believed to be the oldest things in our galaxy – about 12 billion years old. And their orbits all pass through the central region of the galaxy. It is the presence of globular clusters, their age and the fact there is no intergalactic dust and gas that is a major thorn in the side of the big bang theory. It's like this: without the theory of dark matter, etc., the big bang theory predicts an age of the universe at about 9 billion years. However, the globulars are 12 billion years old. It makes it very difficult to say creation occurred nine billion years ago when there are things a lot older right in our own back yard. Then a theory about dark matter appeared, fairly recently, and creation now appears to be about 13.6 billion years ago according to that theory. That is cutting it very close.

---

[1] If we have a sphere with a radius of π and a half-radius of $\pi/2$ then the volume of the inner sphere would be 1/7 of the volume of the outer sphere. That is 0.142857143, which happens to be the measured value of the amount of anti-matter to matter in the universe today. As best we can measure it.

11.1.3 A More Consistent Idea

According to the ages of different parts of the galaxy, there appears to have been an explosion which emitted globular clusters in every direction, but not fast enough for them to escape the mutual gravitational field of their mother proto-galaxy, or whatever gravitational field existed in what was left behind from the explosion. Then appeared a galactic core. Then galactic arms grew out of this core. All of this took billions of years, not many billions, but a few. The stars in the arms, or region I, orbit the centre of the galaxy according to normal gravitational theory, which I will get into later. At the front edge of the arms are regions of stellar formation – areas where stars are being born from the collapse of interstellar dust and gas. Behind the galactic arms are older stars. The stars are generally older farther back from the leading edge of the arms. It appears that there is a disk of interstellar dust and gas. Some disturbance, probably gravitational, is sweeping through the galaxy like a gigantic sprinkler system, and causing the collapse of dust clouds into stars. It should be noted that a cloud of interstellar dust and gas, even though it is gravitationally bound to itself, will not spontaneously collapse into stars. It requires a stimulus, or "seed", to get it to collapse. At first I had thought there was a tidal effect from two black holes at the centre of the galaxy orbiting each other that could cause this effect. However, after checking the work of Dr. Reinhard Genzel in Europe, this does not seem to be the case.

Note: *Dear Mr.Rout, I have attached our most recent research paper on the matter of the mass distribution in the Galactic Center. We see no evidence for a binary black hole. The limits we can place are discussed in the paper. Sincerely, Reinhard Genzel*

(Dr. Genzel has been monitoring and studying the centre of the galaxy for the past 21 years, [Genzel et al (2008)].)

The structure of the galaxy is complex. It's history is not simple. The commonly held theory that the collapse of a huge gas clouds formed galaxies would leave behind an enormous amount of residual gas. It would create a much simpler structure than that presented by galaxies we see today. There is no residual dust and gas between galaxies. Not very little – none, nada, zilch. If there was any, we would not be able to see even the Andromeda Galaxy, our neighbour, let alone many millions of light years. It is highly dubious, practically impossible, for a galaxy to form from the implosion of a cloud of dust and gas. It could not have happened even with the most bizarre cosmological theory. It is much more plausible that it resulted from an outward explosion or expansion from the core. Therefore, I am conjecturing that there must be an upper limit to the mass of a black hole and that limit is in the order of magnitude of the mass of a galaxy. Some time ago, in the past century, a suspicion arose concerning the formation of stars. It came from the question: Why are stars all basically the same mass, given an order of magnitude? We now have to ask the same question of galaxies. They are all roughly the same mass. Why is that? And I don't buy the Finger of God Theory either. I'm with Laplace, [URL-15], on this one.[2]

I must also enter a side note here. From the very latest evidence on the spectacular work of the Event Horizon Telescope Collaboration It is not that a black hole actually explodes as I have postulated. It does appear that galactic arms are forming out of galactic super black holes which are going through a metamorphosis, particularly from what we see of M87. The actual process of a galactic re-birth is more complex than the simple postulation which Werner and I discussed, but we were definitely on the right track.

11.2   How to Make a Galaxy

At the time, some time ago after seeing Werner, this was my idea. I have developed this idea much more in years afterwards. But these were my thoughts back then. Figuring this out is a process. It gets better.

1. feed a black hole with matter forcing it to increase its spin and to cool.

2. reduce rate of feeding black hole as its mass approaches 10^{11} solar masses.

3. when black hole cools to a temperature very close to absolute zero, throw a brick at it and duck.

---

[2] An account of a famous interaction between Laplace and Napoleon is provided by Rouse Ball, [Ball (2010)]: "Laplace went in state to Napoleon to accept a copy of his work, and the following account of the interview is well authenticated, and so characteristic of all the parties concerned that I quote it in full. Someone had told Napoleon that the book contained no mention of the name of God; Napoleon, who was fond of putting embarrassing questions, received it with the remark, 'M. Laplace, they tell me you have written this large book on the system of the universe, and have never even mentioned its Creator.' Laplace, who, though the most supple of politicians, was as stiff as a martyr on every point of his philosophy, drew himself up and answered bluntly, Je n'avais pas besoin de cette hypothèse-là. ['I had no need of that hypothesis.'] Napoleon, greatly amused, told this reply to Lagrange, who exclaimed, Ah! C'est une belle hypothèse; ça explique beaucoup de choses. ['Ah, it is a fine hypothesis; it explains so many things.']"

Homework Assignment #3

1. Derive the formula to determine the temperature of a stationary black hole, having no unbalanced charge, according to its mass.

2. Is there a limit to this temperature?

3. What would be the angular momentum of a spinning black hole?

4. Is there a limit to this angular momentum?

Chapter 12

Galaxies and their Amazing Spiral Structure

12.1 The Dichotomy

There is a dichotomy in philosophy that the search for truth includes the avoidance of falsehood. And modern science is filled with such a high degree of falsehood as to be completely irrational. I am not concentrating on the statement that what the public sees of scientists is the sight of a bunch of raving lunatics, although it appears to be so; I am making the statement that we have been misled by making the wrong assumptions. Much of modern science is directed to support a philosophy that doesn't work. That is why all the work on quantum mechanics and galaxies is so important. It both eradicates falsehood and establishes truth. From this we can determine a philosophy that works.

12.1.1 How Do We Know the Universe Is Not Expanding?

From about 1975, we learn from Misner, Thorne and Wheeler, [Misner et al (1973)], that if the universe was measured to be accelerating outwardly, then there would have to be a dramatic change in our understanding of cosmology, [URL-3].

To put this more clearly, if the universe is measured to be accelerating outwardly, taking time delay of our measurements into account, then the universe cannot possibly be physically expanding. This is because of the cosmological principle that there is no preferred place in the universe and because of the difference between vectors and scalars. A vector's value changes if we change the coordinate system in which the vector is being measured; a scalar's value does not. An example is that if we have a kilogram of crud on the surface of the Earth and we go to the Moon with it, we still have a kilogram of crud. However, if there is a gravitational force of 9.81 Newtons on the crud at the surface of the Earth, we would have about a sixth of that gravitational force on the same crud if it were on the surface of the Moon. That is because gravitational "fields" are just a transformation of coordinate systems. The mass of the crud is a scalar and does not change its value under the change in coordinate systems. The force acting on the crud changes since we have changed the coordinate system by which we are measuring it.

So if we take into account the time delay in measuring the recession speed of distant galaxies and then derive a vector field portraying the rate of expansion at each point in the universe, we see that the magnitude of the vectors in this field vary as we measure each point further out from our position. This shows the accelerated rate at which the universe is measured to be expanding. However, if we take the divergence of this vector field, we see that the divergence is non-zero everywhere in the field with the exception of one particular place, and that is our own location. Because the divergence is a scalar field, it does not change it's value if we change the location where we are taking measurements. This is a property of the divergence. This means that if we ensure that we take time delay of our measurements into consideration and transform all our measurements to some other location, say 200 MegaParsecs away, and re-calculate a new vector field determining the calculated expansion of the universe, we would definitely get a different vector field; however, if we take the divergence of that new vector field, we end up with exactly the same scalar field as before, and even from that point of view, we would see that our own location – here in the Milky Way Galaxy – would be the only location in the universe having a zero divergence. The location of zero divergence is the source of this accelerated expansion. In other words, our Milky Way Galaxy is causing this accelerated universal expansion. Isn't that remarkable? It's the magic of vector calculus and the divergence. Engineers use it all the time.

Now, of course our Galaxy is not causing the universe to accelerate outwardly, it is just that our measurements are showing that. So what in the world is going on? Let's look at something far more simple. Suppose you are sitting in a train at a train station and you suddenly see the station accelerate away from you. It's an illusion, the station is not accelerating away from you, you are accelerating away from the station. Not that many people travel by train any more so it may well be that you have not had that experience. But it is common to have the illusion that the station is moving away from the train when the train begins to leave the station. So if the universe is seen to be accelerating away from us in all directions and it is not actually expanding, then we must be accelerating away from the universe in all directions. How is that possible??? Are we shrinking rather than the universe expanding?? That would be bizarre. First year physics students learn this year after year, it happens to be the result of circular motion. If you are in circular motion, you are constantly accelerating towards the centre of motion. And that makes everything outwardly from the centre of your circular path appear to be accelerating away from you. But, you would be able to tell that because there would be an axis of your circular motion. Say you are on a children's merry-go-round at a playground. And you are in circular motion riding on this merry-go-round. You would see the rest of the playground spinning around you but you could tell that directly above you there would be an axis of rotation – the sky would appear to be spinning about a centre directly above the merry-go-round. We do not see this as a phenomenon in our cosmic measurements. Although I have heard that the universe appears to be spinning at the speed of light by examining the nature of electromagnetic radiation we receive from the cosmos. But the universe is not spinning. We are. But in more than one way. Let us now consider a merry-go-round mounted on a gimbal. In other words, there are different axes of rotation all off-set from each other. And we take some poor unsuspecting soul and put them in the inner merry-go-round and spin them merrily around. Then we have this inner merry-go-round spin on a second merry-go-round which has an axis of spin very much off-axis from the inner one. So the unsuspecting soul is now spinning, in a very convoluted way, about two different rotating axes of rotation. And then we have still yet a third merry-go-round upon which these two are mounted and it is also spinning with an axis of rotation completely off-axis to either of the inner two. Now, inside the head of this poor unsuspecting soul, the surroundings appear to be moving in a completely confused way. It would be very difficult to work out which way everything is spinning. And very definitely, our unsuspecting soul would measure his or her surroundings as accelerating away. If we consider our own axis of rotation, we know the Earth is spinning, so that is circular motion for those on its surface. And the Earth is revolving about the Sun, so that is another axis of rotation and off-set from the axis of rotation of the Earth. On top of that, the Sun is orbiting the Milky way and we may have more complex motions from gravitationally interacting with near-by stars. And, of course, our galaxy could be rotating around the centre of mass between ourselves and the Andromeda Galaxy. And there could be lots of other curved path motions to consider. But all of these accelerations are very small, you may say. And that may well be true, but the effect of bringing in acceleration to our calculations involves multiplying this very small acceleration with very, very large distances. These distances are so large, that we have to account for relativistic effects. Things may be rotating very slowly, however, the acceleration involved has to be multiplied by distance and you have to do the math before you can discount this effect.

As far as the accelerating universe is concerned, there is the cosmological principle that there is no preferred place in the universe. If we take the divergence of an expanding vector field representing an outwardly accelerating universe, then we end up being at the centre of the universe and the source of outward acceleration. And that cannot be. We have known this since 1975. The universe cannot possibly be physically expanding.

12.2   Can We Measure That?

First we need to be able to measure the distances to far-off galaxies. And luckily enough, I know how to do that. I will show you that in a later chapter. Or maybe a number of chapters. Measuring distances to galaxies usually relies on how bright they are – their luminosity. And that is difficult to measure without some pretty advanced instruments. Although today, with digital photography and some off-the-shelf software, you can make some pretty good estimates of galactic luminosity, so I could actually be wrong on that. It does happen. Because of the structure of spiral galaxies and what is making the galaxy have a spiral morphology, I can use the spiral structure of a galaxy and its maximum speed of spin in order to find its distance. And I can delve into the databases to get the red shift of the galaxy itself. Red shift is a measure of the displacement of a spectral line. When an atom gets hit with light an electron may absorb it and jump to a higher energy level as a result. Then it may very quickly release this energy and drop down to a lower energy level. Each energy level is a certain amount and we have measured many of these energy levels. These levels depend on the material whose atoms and electrons are absorbing and releasing these discreet energy levels. But when the energy is released in a photon, the photon may well head off in a different direction than the direction of the absorbed photon. So light may be emitting from a source, and we see all the light that is not absorbed, but the light that is absorbed is scattered by electrons in some intervening material and shows up as a dark lines in the spectrum. An example is the Balmer line in Hydrogen at 656.45377 nano-meters having a corresponding frequency of 456684798.992014 MegaHertz. If things are moving away from us, this spectral line appears to be shifted towards the red end of the spectrum. The difference in measured frequency between the moving scattering material and us is called a redshift. If material scattering light is coming towards us, it would be blue shifted. Also by measuring the difference between the shifted spectral line and what we see from non-moving material in a laboratory, we can measure how fast things are moving away or towards us. But there are other ways that light can be red shifted or blue shifted. Light can bounce off of something and its frequency would red shift. Light can pass though curved space-time and it would be red shifted. And light emitted from a spinning object is blue shifted. This last statement is of most interest to us in this discussion. That is because we are not concerned with the blue shifting of light from something that is spinning, because we are the ones spinning. If you are spinning, light seen by you would be red shifted. Light emitted from a gravitational well is red shifted. If you are in the gravitational well, light falling inwardly would be blue shifted. So if you are in a spinning environment, light from the cosmos would appear red shifted and the universe would appear to be accelerating away from us.

But can we measure it?

Well, sort of. It involves a lot of math. I'll do that later. But for now, let me explain where I am so far. I have to find the location and distance to each galaxy. Then I have to calculate how fast that galaxy appears to be spinning around us because of the rotation of the Earth, then do the same to remove the effects from the Earth going around the Sun and finally apply that to the red shift of the galaxy being measured. Unfortunately, if that is done, we end up with galaxies that appear to be moving away from us at speeds far greater than the speed of light. And we cannot allow that. Some mental gymnastics are involved. We have to bring in the effects of relativity and rotating reference frames. We have work to do.

12.3   The Spiral Structure
of NGC 3198

It was the discovery of the spiral structure of NGC 3198 and resultant analytical model of galaxies that has formed a huge breakthrough for me in the discovery of the knowledge of everything. My Daughter Tara found out when she was 12 that I didn't know everything and she has been so angry with me as a result. This requires some explanation. You know how it is that our parents lied to us? It happens to everyone. We grow up and at some point, when we become adults, we find out that our parents lied to us. We all go through it. Well, I knew that my children would also go through it. I figured that I should tell them a deliberate lie so that when they hit some semblance of adulthood, they would see it as funny, not take the shock of it too hard and get on with life not trusting anyone and figure things out for themselves. So, every time my kids asked me something like: "Why is the sky blue?" I could answer: "Because the absorption coefficient of the atmosphere varies as λ to the fifth." And they

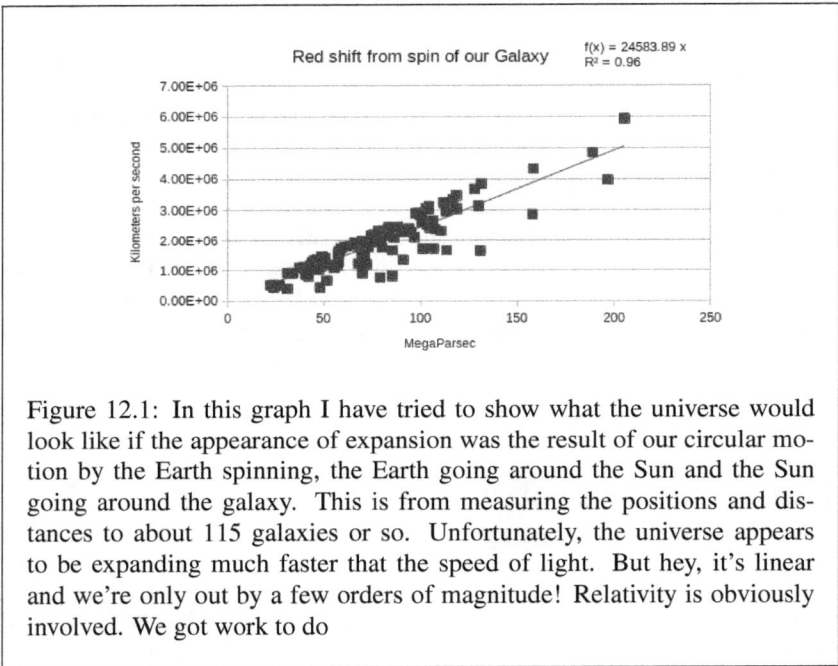

Figure 12.1: In this graph I have tried to show what the universe would look like if the appearance of expansion was the result of our circular motion by the Earth spinning, the Earth going around the Sun and the Sun going around the galaxy. This is from measuring the positions and distances to about 115 galaxies or so. Unfortunately, the universe appears to be expanding much faster that the speed of light. But hey, it's linear and we're only out by a few orders of magnitude! Relativity is obviously involved. We got work to do

would then ask: "How do you know that?" And I would answer that because I am a father, I have to know everything. It has taken be quite some time to do so, however, I finally did it and now I know everything. Of course, when any of them caught me making a mistake, I just said: "I knew that. I just forgot. I know everything, I just can't remember it all." They figured it out pretty quickly. However, Tara is very special. We used to talk about infinity, eternity and how to add proper fractions in your head. She was three. Then she found out I had lied and didn't know everything. She was devastated. She was only 12 when she found out. So, now I actually have to learn and know everything to make it up to her. This treatise is a work of love for her. She taught me about infinity and true genius. I didn't think there was such a thing as true genius until I had deep discussions about infinity and adding proper fractions when she was three. I work this hard at learning everything just to try and keep ahead of her. I have three kids, and keeping ahead of any of them is a full time job.

I was on an internet chess team, the Amateur Astronomy Chess Team, when one of the players posted a note asking what we thought dark matter was. I said it doesn't exist. I was then asked to explain the flat velocity profile of galaxies. I said, "What do you mean?" The questioner said that the stars all go around the galaxy with the same tangential velocity. I said I didn't believe it. He sent me the data. Sure enough, the stars go around the galaxy at the same speed. Stars closer to the centre go around slower, but a little ways out from the centre, all the stars go around the centre of the galaxy at the same tangential velocity. So I did a fundamental straight forward analysis of the gravitational shape of what is causing this.

First, we have the fundamental equation of circular motion:

$$\frac{v^2}{r} = a$$

where v is the tangential velocity of some object in circular motion about some centre. r is the distance from the object to the centre and a is the centripetal acceleration the object experiences as a result of travelling in circular motion. Note that v is a constant in this scenario. Since we are dealing with some object, let's say it has a mass m. We multiply both sides by m to yield:

$$m\frac{v^2}{r} = ma$$

and, as we all know, $F = ma$ so we stick in force.

$$m\frac{v^2}{r} = F_g$$

I have put in F_g to denote gravitational force pulling the object into a circular orbit about some centre. I feel this is reasonable since I know the force acting on the object is neither electromagnetic nor a strong nuclear binding force. Now, according to my friend Newton, gravitational forces are inversely square and are attracted to the centre of mass of different bodies. So we treat our object as a body being attracted to a collection of stuff which acts as though it has a centre of mass at the centre of the orbital motion of our object. This way we can reduce the problem to a two-body problem with our object in circular motion about the centre of mass of other bodies and that centre of mass would be calculated as having the mass required to keep our object in circular motion. Let me call that required mass at the centre of orbit as an "effective mass". Since this effective mass at the centre of orbit could possibly vary in its influence on our object, depending on the object's distance from the centre, we can say there exists a function, $M(r)$ which is the effective mass at the centre of orbit which depends on the distance, r, from the centre to our object in order for the object to be in circular orbit. (It's a cheat, but a valid one). So from Newton we have:

$$F_g = G\frac{mM(r)}{r^2}$$

plugging this into the stuff above we have:

$$m\frac{v^2}{r} = G\frac{mM(r)}{r^2}$$

and hacking around and rearranging we have

$$\left(\frac{v^2}{G}\right) r = M(r)$$

and since v and G are constants we have a linear equation. In other words, galaxies are sticks. They behave as though they have a linear orientation of matter, and overall, the galaxy has a constant linear density. It is the only way that the tangential velocity of all the material can be constant. And, of course, this makes no sense at all. If it's a stick, the material would orbit faster and faster further out from the centre like a rigid body. But it can't rotate like a rigid body because, to put it bluntly, it doesn't, and yet, because the attractive force, a la gravity by Newton, it must. It's gotta be a stick. And on top of all that, galaxies don't look like sticks, they look like spirals ... and that is where Minkowski, Lorentz and Einstein step onto the playing field.

12.4   Kepler and Dark Matter

Present explanations for the behaviour of the orbital revolution of stars in galaxies involve the hypothesized presence of an exotic substance known as dark matter. A very early mention of the existence of dark matter was made by Zwicky, [Zwicky (1937)], as a result of observing nebulousities which are now known as galaxies. However, at that time, Zwicky did not give any definition or detailed properties of this substance, nor suggest it was anything exotic. The term "dark matter" was mentioned simply as matter that is dark or non-luminous. He also points out that he could not explain the luminosity distribution of galactic nebulae by relating it to the calculated mass distribution using the assumptions of a Keplerian system. We shall refer to this relationship between mass and luminous material as the mass-luminosity relationship. Dark matter was first described as simply non-luminous matter by Zwicky, Rubin et al., [Rubin & Ford (1970)], in their measurements of the tangential velocities of stars in orbit about M31. They discovered a peculiar relationship, which has since risen to the suspicion that this dark material may have certain exotic properties yet to be defined. This relationship has been seconded, notably, by Mathewson, [Mathewson et al (1992)], Begeman, [Begeman (1989)] and Persic, [Persic & Salucci (1995)], as well as many others. A rotation curve for a galaxy is a graph of the measured tangential velocity of stars and material about the centre of the galaxy vs. the angular distance, in arcsec or arcmin, of the stars and material from the centre of the galaxy. These measurements are based on the Doppler shifts of these stars, the red shift of the galaxy itself and the galaxy's angle of incline to the celestial sphere. We show an example in Figure 12.10 along with other examples in Figure 12.11. These figures are described in detail later. Rotation curves show a distinctive relationship between tangential velocity and radial distance to the centre of the galaxy. The graph grows linearly from the centre of the galaxy and then flattens in the outer regions. We denote this distinctive curve as the flat velocity rotation curve of galaxies. A cursory examination of the expected rotation curve of a Keplerian system and the flat velocity rotation curve observed, leads to the conclusion that galaxies either portray inexplicable behaviour or that galaxies do not meet the requirements of a Keplerian system and yet have explicable behaviour using some other model or system.

Kepler, [Kepler (1619)], determined certain laws concerning the orbits of planets around the Sun. Notably that the orbits were elliptical and that the period of the planet's orbit varied cubically as its distance from the sun squared. Newton, [Newton (1687)], in turn, determined that this discovery was the result of the sun's gravitational field obeying an inverse square relationship. A Keplerian system consists of a central massive region, i.e. the Sun, surrounded by light objects, i.e. planets, which do not have sufficient masses themselves to effect the overall system. Apart from small orbital perturbations as a result of neighbouring planets, this model has stood the test of time for more than 350 years. A Keplerian model has a very important requirement; it must consist of an overwhelmingly massive central body or region so that the gravitational fields of orbiting material or objects can be discounted and ignored. For galaxies this would mean that the stars within the galaxy would not have any gravitational influence on each other. Obviously stars interact with each other in a galaxy. Therefore a Keplerian model is inappropriate to use to describe the structure and behaviour of a galaxy

Because of the apparent discrepancy between Keplerian gravitational theory and observations of rotational velocity profiles of galaxies, we investigated, using the general theory of relativity, [Einstein et al (1923)], to resolve this discrepancy and we derived an analytic galactic model as a result. The consequences of this model were then compared to other galactic parameters in order to validate the model. These parameters include the distances to galaxies, their luminosity profiles and masses. We have also attempted to compare the lengths of bars in barred spirals to that predicted by the model in order to add to its verification. We begin a presentation of this investigation by examining the rotation curve of NGC 3198.

What follows is a formal solution ...

12.5 An Analytic Mathematical Model to Explain the Spiral Structure and Rotation Curve of NGC 3198

by Bruce Rout and Cameron Rout

12.5.1 Abstract

PACS:98.62.-g An analytical model of galactic morphology is presented. This model presents resolutions to two inter-related parameters of spiral galaxies: one being the flat velocity rotation profile and the other being the spiral morphology of such galaxies. This model is a mathematical transformation dictated by the general theory of relativity applied to a rotating polar coordinate system that conserves the metric. The model shows that the flat velocity rotation profile and spiral shape of certain galaxies are both products of the general theory.

Validation of the model is presented by application to 878 rotation curves provided by Salucci, and by comparing the results of a derived distance modulus to those using Cepheid variables, water masers and Tully-Fisher calculations. The model suggests means of determining galactic linear density, mass and angular momentum. We also show that the morphology of NGC 3198 is congruent to the geodesic within a rotating reference frame and is therefore gravitationally viscous and self bound.

12.5.2 Introduction

An examination of previous studies of galactic rotation curves and morphology shows that, although relativistic effects of accelerating reference frames have been investigated, no integral resolution to theoretical discrepancies has been found. The special relativistic effects of material in orbit about the center of a galaxy appear to be negligible since the tangential velocity of such stars have been measured as moving at non-relativistic speeds. Nevertheless, it can be shown that general relativistic effects, as a result of centripetal acceleration, are significant and mask the measure of tangential velocity. Tangential velocities of stars in galaxies are measured using the shifting of spectral lines. This shifting is assumed to be strictly a Dopplerian effect. However, general, as well as special, relativistic analysis show that the shifting of spectral lines within rotating bodies are affected by both Doppler shift and effects of centripetal acceleration.

Historically, Keplerian rotational dynamics have been assumed in examining spiral galaxies. The observed tangential velocity of matter does not match a Keplerian model. This has resulted in the inference of significant amounts of non-luminous matter being required for a Keplerian model to match observed orbital behaviour of galaxies as portrayed in rotation curves. To illustrate, an example of a rotation curve can be seen in Figure 12.2 which shows work by Begeman, [Begeman (1989)] using the luminosity curve of NGC 3198 to calculate the expected rotation profile assuming Keplerian dynamics in comparison with measured values.

Two propositions which have resulted from the discrepancy between the expected and the observed rotation curves of galaxies such as NGC 3198 are: first, that more mass must exist within the system than appears in order for Keplerian dynamics to apply; second, that some new hypothesis on the laws of gravitational dynamics exists in lieu of Kepler's laws, [Kepler (1619)]. This paper demonstrates that existing scientific theories can explain the orbital behaviour of galaxies without requiring assumptions of either additional mass or undiscovered gravitational principles. This is done by refuting that galaxies behave as though they consist of non-interacting particles of zero viscosity orbiting a central massive body; rather, that they consist of interacting orbiting bodies to which relativistic considerations must be applied. A resolution is presented here in the form of an analytical model which is a mathematical spiral having a flat rotation profile resulting from the application of Lorentz transformations in an accelerating environment.

Some Previous Approaches

A comprehensive overview examining the spiral structure of galaxies was done by Binney, [Binney & Tremain (2008)]. In this work, Binney assumes a non-viscous Keplerian model, provides extensive substantiation that spiral galaxies are indeed spirals at all wavelengths, and laments the lack of a complete exposition: "despite much progress, astronomers are still groping towards this goal," he writes. He presents a model of spiral morphology of galaxies based on a proposition of tidal forces generated by density waves. He rejects a model of stationary spiral structure and utilizes a rotating coordinate system. We present here a model in which a Keplerian non-viscous assumption is replaced by a general relativistic, highly viscous model also utilizing a rotating coordinate system. Furthermore, a similar consequence is found in which the metric, resulting from a rotating polar coordinate system, rather than density waves, creates a model having a stable rotating spiral structure of material.

Cooney, [Cooney et al (2012)] attempted to resolve the discrepancy between Keplerian motion and observed line width profiles through a general relativistic approach that used the Schwarzschild solution to constrain the metric with some success. Menzies, [Menzies & Mathews (2006)] investigated and criticized a different model presented by Cooperstock, [Cooperstock & Tieu (2005)] which, along with Gallo, [Gallo & Feng (2010)], utilized gravitational fields and associated curvature in the field equations balanced against Dopplerian Lorentz transformations in order to obtain flat velocity curves of galaxies both numerically and analytically. Menzies showed this approach resulted in requiring an infinite amount of mass. These previous approaches show that the study of galactic structure and underlying physical models is still ongoing.

Galaxies as Non-Keplerian Systems

To apply the laws dictated by Kepler, the system must behave as a central massive region around which particles orbit without significantly interacting with each other, such as in the solar system. A useful description for such a system is provided by Zwicky as having negligible gravitational viscosity, or "zero-viscosity". If the distribution of matter in a galaxy were such that the gravitational viscosity was not negligible, but rather high enough to "equalize the angular velocity throughout such systems regardless of the distribution of mass", [Zwicky (1937)], then such a galaxy would no longer be comprised of a central disc rotating with orbiting zero-viscosity matter.

Galaxies have a bright, dense central region with a sparse outer disc. Such a luminosity distribution suggests galaxies should behave as a zero-viscosity Keplerian system as modeled by Begeman (see Figure 12.2). However, observations by Begeman, Mathewson, [Mathewson et al (1992)], and Persic, [Persic & Salucci (1995)], have shown that orbital velocities do not behave accordingly. We shall show that this discrepancy can be resolved by applying general relativistic effects and comparing it to observation. It is important to note that special relativistic effects apply to inertial reference frames and general relativistic effects apply to accelerating reference frames which include rotating bodies. This is especially relevant to deriving an analytical model of galaxies because general relativity, rather than special relativity alone, must be applied to rotating bodies since they are not inertial reference frames. For example, measuring the shifting of spectral lines in a heavily curved space, such as in the proximity of a black hole, must take into account the shifting of spectral lines independently of velocity. Similarly, space can be heavily curved in a galaxy, due to centripetal acceleration, which is also independent of the measured velocity.

12.5.3   Relativistic Galactic Model

Restricting ourselves to a mathematical approach, wherein we retain the constancy of measures of the speed of light, a model is derived using a measure in four dimensions. Comparisons can then be made of measures of length and time between coordinate systems which are moving and accelerating relative to each other. Consider two four-dimensional Minkowski spaces in which exist standard clocks and rulers at all points. We may compare the behaviour of these clocks and rulers through various transformations which conserve the metric, keeping the measure of the speed of light constant through each transformation. We may then transform measures into an appropriate coordinate system which shall prove useful in determining properties of spiral galaxies.

We derive the metric of a rotating body from Einstein, [Einstein et al (1923)], by considering the shape of a geodesic in a rotating system using Lorentz transformations.

Two important properties of a radial geodesic in a rotating coordinate system are as follows:

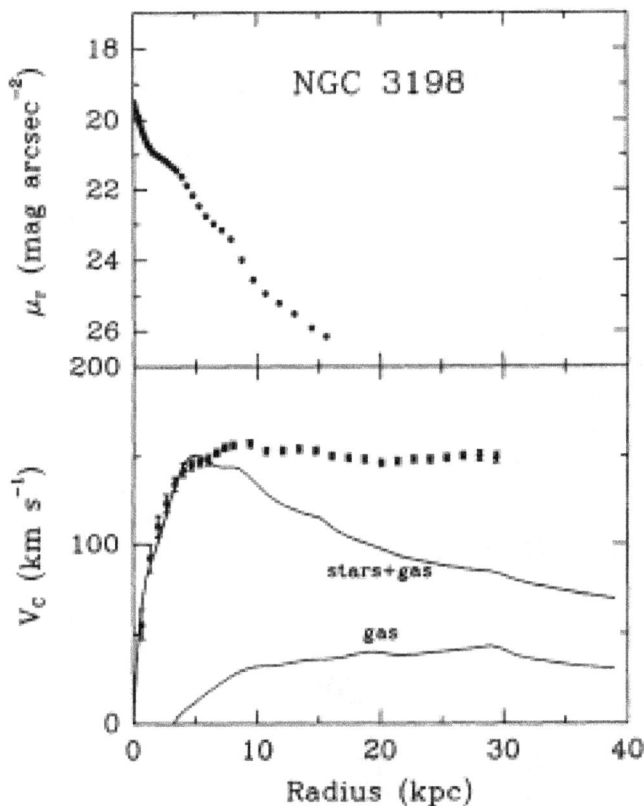

Figure 12.2: Begeman's plot of observed rotation velocities (bottom) compared with rotation curve predicted from the photomectric data (top) assuming a constant mass-luminosity ratio and z-thickness. Begeman used a sech-squared law with disk thickness of 0.2 × the disk scale length and included the contribution of the gas component.

Firstly, the tangential velocity of such a coordinate system behaves peculiarly as a function of distance from the center. A relativistic model must take into account the fact that the tangential speed of a rotating coordinate system must never reach the speed of light. It is shown that the measure of tangential velocity reaches a limit as distance from the center of rotation increases.

Secondly, the path of light travelling radially outward from the center of rotation traces a spiral-shaped path within the rotating coordinate system. More specifically, it approaches an Archimedes' Spiral, defined as a spiral having a constant pitch. The pitch, κ, of an Archimedes' spiral in polar coordinates is analogous to the slope of a straight line in Cartesian coordinates. That is:

$$r = \kappa\theta \tag{12.1}$$

as compared to

$$y = mx \tag{12.2}$$

where r is radial distance from the origin, θ is the angle subtended anti-clockwise from the x-axis, κ is a constant; x and y are the usual x-y coordinates and m is the slope of a straight line – a constant. These equations are well known and are presented in the general sense to show similarity in form.

We first show the effect of a strictly mathematical transformation of pixels of a digital image from one coordinate system to another. Following this, we examine the effects of general relativity in rotating coordinate systems which result in these transformations.

A Spiral Transformation

We present an analytic model in which the distinct spiral shape of galaxies appears asymptotically as a function of distance from the center. Following which, we show how it can determine certain galactic parameters which are of interest. Please note that the resultant model of spiral galaxies is not a pure Archimedes' Spiral but uses the morphology of an Archimedes' Spiral as an asymptote to which the shape of spiral galaxies quickly approaches as a function of radial distance depending on the galaxy's rate of rotation. This model is applicable to galaxies which have a distinct flat rotation profile; galaxies which do not have flat velocity profiles would deviate from this model.

The following equations determine a transformation which is applied to the pixels of an elongated blob as in Figure 12.3 (left).

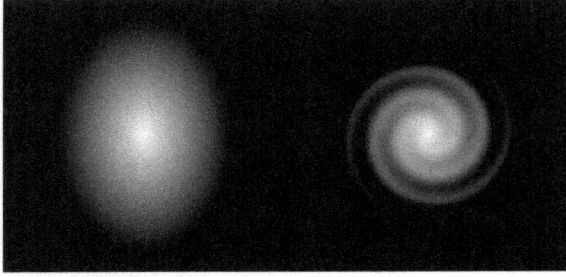

Figure 12.3: The left figure is an elongated blob of white against a black background in Cartesian coordinates with the origin at the center of the figure while the figure on the right is the transformation of the left figure using Equations (12.3) and then portrayed on an orthogonal rectilinear grid. The transformation used a value of $\kappa = 20$.

Consider the following spiral transformation:

$$\begin{aligned}
r &= \sqrt{x^2 + y^2} \\
\theta &= \arctan(y/x) \\
r' &= r \\
\theta' &= \theta - r/\kappa \\
x' &= r' \cos(\theta') \\
y' &= r' \sin(\theta')
\end{aligned} \tag{12.3}$$

where x and y are the coordinates of a particular pixel in the originating figure with the origin at the center.

The first two equations in the transformation of Equations (12.3) transform the pixels of a figure into polar coordinates. The second two transform the pixels onto a spiral rotation while the last two transform the results into Cartesian coordinates and can be portrayed as rows and columns of a resultant figure. Note that Figure 12.3 (right), is a shape resembling a spiral galaxy. This mathematical transformation results in a figure which shows a distinct spiral morphology even though the original figure is somewhat amorphous. The units used in the transformation are pixel-length units with the origin of both polar and Cartesian coordinates at the centre of the blob and spiral shape respectively. A pixel smoothing software was used following the transformation to eliminate gaps in the resultant spiral and the pixel brightness following the transformation is therefore slightly altered.

70

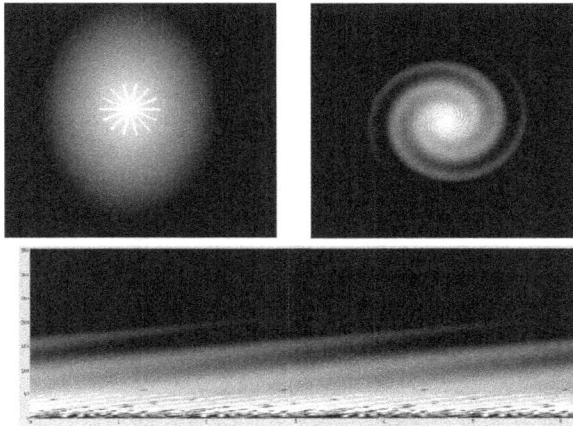

Figure 12.4: The top left figure is an elongated blob with bright radial structure at the center. This blob is then transformed with a rotation in the upper right figure. The upper right figure is then portrayed in rectilinear coordinates with r as the ordinate and θ the abscissa and shown in the lower figure. The lines in the lower figure slope to the right as a result of the clockwise "spin" of the upper right figure.

The superimposition of a star shape at the center of an oval cloud is shown in Figure 12.4 (upper left). The spiral transformation described by Equations (12.3) is applied to the pixels in this figure and then shown in Figure 12.4 (upper right). This figure also has a clearly defined spiral morphology including the brighter, smaller spirals in the middle of the figure. Figure 12.4 (lower) is a portrayal of the upper right figure with the ordinate as r and abscissa as θ. The series of parallel straight lines and their slope denote a value of $\kappa = 20$, which was used in the transformation.

This is a very "powerful" transformation in that an original figure, which may resemble nothing more than something akin to a slightly elongated blob, resembles a distinct spiral following this transformation. As a matter of fact, we could take a picture of a moose and put it through the above transformation and the moose would then resemble a spiral galaxy. If the moose was large enough, as a rotating body, it would have a flat velocity rotation profile. The analytic model which follows shows that the general theory of relativity, due to the acceleration as a result of circular motion, compels this morphology. Therefore, the mathematical transformation demanded by the general theory of relativity, rather than the morphology of material being so transformed, results in both the

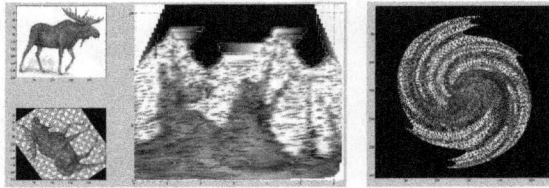

(a) A moose and r vs theta transformation   (b) A moose galaxy

Figure 12.5: On the left is a picture of a typical majestic Canadian moose and the moose transformed into r vs theta as per the first half of equations (12.3). On the right is the same moose transformed through a spiral transformation resulting in a morphology typical of a spiral galaxy with a high degree of flocculence.

spiral morphology of certain galaxies and their flat rotational velocity profile. This is demonstrated in the following sections.

Comparison of the Spiral Transformation to Some Spiral Galaxies

For a given spiral such as in Figure 12.4 (upper right), consider mapping r onto y and θ onto x , in order to observe a possible linear orientation. This results in a measure of κ, equal to the slope of bright parallel lines in Figure 12.4 (lower).

Continuing, κ was measured for three different galaxies as shown in Figures 12.6-12.8. In these figures, the position of each photograph's pixel is transformed from a row-column coordinate to polar coordinates in which the center of the galaxy is the origin, r is the distance in pixels from the center of the galaxy and θ is the angular measure from a horizontal axis as in polar coordinates. The position of these pixels are then transformed where r is the ordinate and θ the abscissa. In these figures, Equation 12.1 is investigated by inspecting the linear orientation of the resultant pixel greyscales.

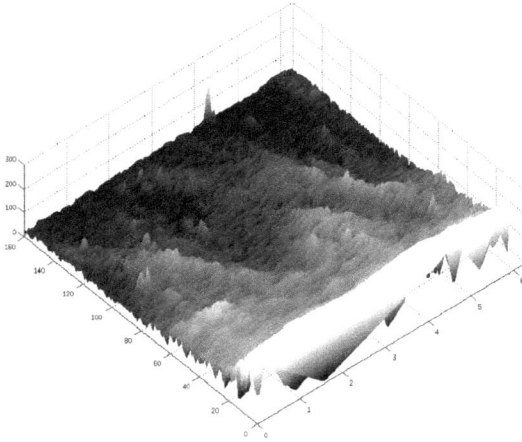

Figure 12.6: A three dimensional luminosity figure of NGC 4321 where θ is the abscissa and radial distance from the center of the galaxy is the ordinate. Note the linear orientation of luminosity elevations in the figure which correspond to the spiral arms of the galaxy. Also note the spate of flocculence near the center of the galaxy is also oriented linearly with the same slope. The linear ridges appear parallel and encourage the derivation of Equation (12.20) from physical parameters.

Note the three dimensional representation of luminosity vs. r and θ of NGC 4321 in Figure 12.6. In this figure there are two obvious linear ridges of greater luminosity oriented with consistent negative slope emanating from the abscissa and separated by π radians. Also in this figure there are other shorter ridges and peaks emanating from the abscissa with similar slope. In Figure 12.9, also portraying NGC 4321, κ is approximated to be -32 pixels per radian which is the value of the slope of the two lines superimposed on the figure.

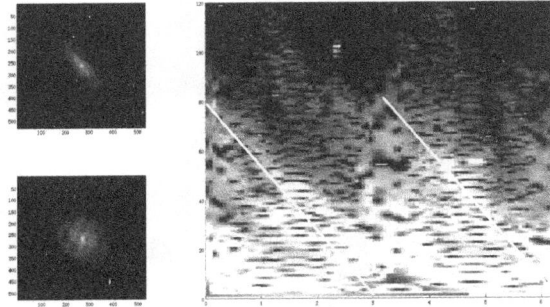

Figure 12.7: This is a portrayal of a digital photograph of NGC 3198 with the original photograph on the upper left, a transformation of the pixels of the galaxy so that the galaxy appears as viewed from directly above on the lower left and a transformation of the lower left photograph, transforming the pixels of the photograph onto a plot of radial distance from the center of the galaxy vs. θ as per polar coordinates, on the right. Note in the transformed photograph on the right, the two bright linear orientations of lighter shades are parallel and a horizontal distance of π radians from each other with a slope of -26 pixels per radian.

The galaxy NGC 4321 is oriented with a small angle of incline. As a result, the above described transformation can be conducted without requiring alteration, which is portrayed in Figures 12.6 and 12.9. However, we examine NGC 3198 in Figure 12.7 by first correcting for the angle of incline as shown in the figures on the left and then applying the above described transformation, the results of which are shown on the right of the figure. From this figure we can estimate a value for κ of -26 pixels per radian.

IC 239 is also a distinct spiral galaxy with a small angle of incline. Figure 12.8 is a result of the above transformation in which a value of -11 pixels per radian can be estimated as a value of κ in a spiral equation approximating the distribution of luminosity for this galaxy.

These observed properties can be explained through general relativistic considerations.

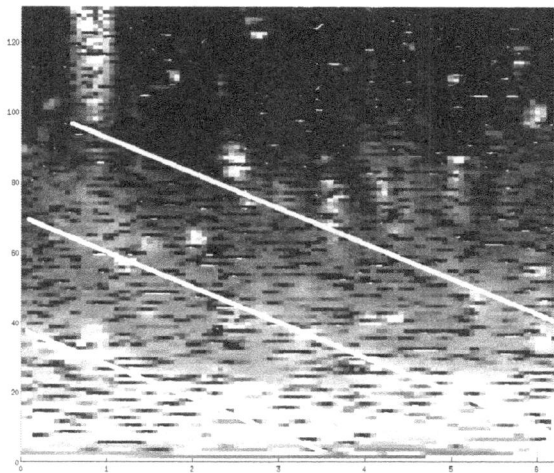

Figure 12.8: This is a portrayal of a digital photograph of IC 239 as a result of transforming the pixels of the photograph onto a plot of radial distance from the center of the galaxy vs. θ (as per polar coordinates). Note the prominent linear orientations of bright pixels are parallel and a horizontal distance of π radians from each other. These lines have a common slope of approximately -11 pixels per radian. The line on the far left continues onto the right. The lines overlay areas of greater luminosity and mark the positions of the two spiral arms emanating from the center of the galaxy.

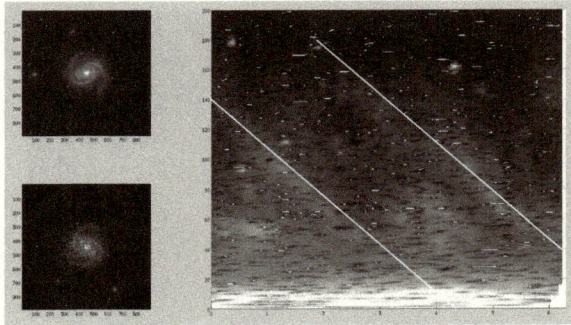

Figure 12.9: This is a portrayal of a digital photograph of NGC 4321 (upper left), a rotation of the photograph (lower left) and the result of transforming the pixels of the photograph onto a plot of radial distance from the center of the galaxy vs. θ as per polar coordinates (right). Note the two distinct bright linear orientations of higher luminosity which have been overlaid by straight lines in the figure on the right. These lines are parallel and a horizontal distance of π radians apart and have a common slope of -32 pixels per radian.

Lorentz Transformations in a Rotating Coordinate System

The Lorentz factor between two coordinate systems moving with instantaneous velocity, v, relative to each other is:

$$\gamma = \frac{1}{\sqrt{1 - v^2/c^2}}. \tag{12.4}$$

If v is a constant then the Lorentz factor can be applied according to well-known special relativity. However, if two coordinate systems are accelerating relative to each other, then v is not a constant and the Lorentz factor is a variable. In such a case, equation (12.4) can only be applied for an instant.

The model presented uses the coordinate systems, reference frames, and application of Lorentz foreshortening in a rotating system as described by Einstein, in the following quote:

In a space which is free of gravitational fields we introduce a Galilean system of reference $K(x, y, z, t)$, and also a system of co-ordinates

76

$K'(x', y', z', t')$ in uniform rotation relatively to K. Let the origins of both systems, as well as their axes of Z permanently coincide. For reasons of symmetry it is clear that a circle around the origin in the X, Y plane of K may at the same time be regarded as a circle in the X', Y' plane of K'. We suppose that the circumference and diameter of this circle have been measured with a unit measure infinitely small compared with the radius, and that we have the quotient of the two results. If this experiment were performed with a measuring-rod at rest relatively to the Galilean system K then the quotient would be π. With a measuring rod at rest relatively to K', the quotient would be greater than π. This is readily understood if we envisage the whole process of measuring from the "stationary" system K, and take into consideration that the measuring-rod applied to the periphery undergoes a Lorentzian contraction, while the one applied along the radius does not, [Einstein et al (1923)].

Consider two polar coordinate systems having coincident origins, one is not rotating and is denoted as the K system, the other is rotating with some angular velocity ω_0 as measured by K and is denoted as K'. We transform the coordinates of the K' system into the coordinates of the K system.

The influence of gravity follows a geodesic. Thus, the shape of a geodesic can be transformed from K' to K.

Consider further the behavior of two identical clocks, one in the K system and the other rotating in the K' system at coincident points. The clock in the K' system will be a slower clock as measured by the one in the K system. This effect is more pronounced at other coincident points further from the origin of the two systems. Therefore the Lorentz factor describing this effect will not be a constant throughout the systems and will vary as a function of r, the distance from the origin. Since the Lorentz factor is not a constant, General Relativistic effects, rather than Special Relativistic effects, need to be applied.

Because clocks in K and K' measure the passage of time differently, the period of rotation of the K' system will be different than that measured by clocks in the K system. Although all clocks in the K system will measure the K' period of rotation the same, say some value T_0, the clocks in the K' system will measure variable periods of rotation and this period will vary as a function of r due to the differing tangential velocities of points in K' also being a function of r. Let us denote the period of revolution of K' as measured by clocks in K' as $T'(r)$. The ratio of $T_0/T'(r)$ is the Lorentz factor by definition and we shall denote this variable Lorentz factor as $\gamma(r)$.

It follows that,

$$\gamma(r) = \frac{T_0}{T'(r)} \tag{12.5}$$

as above. If we denote the angular velocity as ω_0 as measured in K and $\omega'(r)$ as measured in K', we therefore have:

$$\omega'(r) = \frac{2\pi}{T'(r)}$$

$$\omega_0 = \frac{2\pi}{T_0} \tag{12.6}$$

$$\omega'(r) = \frac{\omega_0}{\gamma(r)}.$$

Please note that identical rulers differ in their measures in the two systems. Rulers in K' will be foreshortened in the tangential direction relative to K. The rulers will be identical if oriented in the radial direction. Of course the measure of the tangential velocity in both systems would be identical at coincident points provided one was consistent in using the same clocks and rulers restricted to the system in which the tangential velocity is being measured. Here also, care must be taken to define tangential velocity as arc length traversed over time. It will be shown that the variation of the Lorentz factor, based on differences in the measure of angular velocities, that is $\omega'(r)$ and ω_0, will affect the difference in measures of shifts in spectral lines. This effect could possibly be compared to that of curvature, however, in the opposite sense of curvature described by the Schwartzchild metric in gravitational fields. In a gravitational field, the Lorenz factor increases closer to the source of the field, in a rotating system such as K', the Lorenz factor increases further from, rather than closer to, the center of rotation.

Returning to our presentation of transformations between K' and K, please consider the following well known equation for circular motion in general:

$$v = \omega r. \tag{12.7}$$

where we usually consider v as tangential velocity in a rotating system, ω as the angular velocity, and r as radial distance. This is a general case without considering relativistic effects and well known since the time of Galileo. However, in the case of a rigid body, ω is constant; in the case of a rotating coordinate system which conserves the metric, the angular velocity is a variable and is denoted here as $\omega'(r)$. As soon as we bring in the effects of the differences in clocks and rulers between K' and K, we deviate from classical, or Newtonian, rotational dynamics. Care must be taken in the transformations from one system to the other.

Consider some distance r from the coincident origin in both systems as a parameter. At each parametric r, and for some very small moment of time, at two coincident points in K' and K some distance r from the coincident origin,

$$v'(r) = \omega'(r)r, \tag{12.8}$$

where $v'(r)$ and $\omega'(r)$ are tangential velocity and angular velocity as measured in K'; then, parametrically,

$$\gamma(r) = \frac{1}{\sqrt{1 - v'(r)^2/c^2}}. \tag{12.9}$$

Then,

$$\gamma(r) = \frac{1}{\sqrt{1 - \omega'(r)^2 r^2/c^2}}. \tag{12.10}$$

Substituting equation (12.6) we have:

$$\gamma(r) = \frac{1}{\sqrt{1 - \omega_0^2 r^2/(\gamma(r)^2 c^2)}}. \tag{12.11}$$

Solving for $\gamma(r)$ we have:

$$\gamma(r) = \sqrt{1 + \frac{\omega_0^2 r^2}{c^2}}. \tag{12.12}$$

This yields the spatially two-dimensional time dependent metric in K',

$$ds^2 = \frac{c^2}{\gamma(r)^2} dt^2 - dr^2 - \gamma(r)^2 r^2 d\theta^2. \tag{12.13}$$

We have used

$$-g_{tt}^{-1} = g_{\theta\theta} = \left(1 + \frac{\omega_0^2 r^2}{c^2}\right) \tag{12.14}$$

where $g_{\mu\nu}$ is the metric tensor in order to derive equation (12.13).

Equation of a Spiral Geodesic

The path of a photon travelling radially outward from the origin would travel in a straight line according to coordinates in K but would trace out an Archimedes' Spiral within K'. Furthermore, as the photon travels outwardly it passes over sections of K' whose local clocks and tangential distance measures deviate from

79

those measured within K, according to the Lorentz factor as described in equation (12.12).

We see from equation (12.12) that a linear relationship between $\gamma(r)$ and r begins to be established asymptotically at distances $r > \frac{c}{\omega_0}$ from the center of the rotating system. As a result, the calculated tangential velocity using differences in spectral lines, which is effected by $\gamma(r)$ would approach an asymptote which we denote as v_{max}.

Discussion of Units

In examining the metric of equation (12.13), and the equation for the Lorentz factor in equation (12.4), we see an interchangeability between time and space coordinates by either multiplying the time coordinate by the speed of light or by dividing spatial coordinates by the same value. Coordinates in a Minkowski space are (ict, x, y, z). Converting from a Minkowski space in MKS units to completely unit-less dimensions such as $\tilde{R}$ for radial distance, $\tilde{v}_{max}$ for the asymptote of tangential velocity and $\tilde{\omega}_0$ for angular velocity we have:

$$\tilde{v}_{max} = \frac{v_{max}}{c}, \tag{12.15}$$

$$\tilde{r} = \frac{r}{c\tau}, \tag{12.16}$$

and

$$\tilde{\omega}_0 = \omega_0 \tau \tag{12.17}$$

where τ is the number of seconds in a year, assuming the original measurements are in MKS units and radians are considered unit-less.

For convenience one may consider $\tilde{r}$ as the value of spacial measurement in ly and $\tilde{\omega}_0$, as a measure in radians per year.

The law of rigid bodies predicts a constant angular velocity throughout the entire rotating body. However, taking relativity into account, the measure of angular velocity varies with radial distance and yet the entire body appears rigid, or has a stable shape over time. The value of $\tilde{\omega}_0$ is constant throughout the rotating coordinate system: however, the value of angular velocity, $\omega'(r)$, is not. As a result, the rotating coordinate system, although appearing as a rigid rotating spiral, would have different measures of $\omega'(r)$ at various radial distances. This is contrary to the laws of a rigid body according to Newton's laws of motion or to Lagrangian mechanics which are based on non-relativistic considerations.

In general, the Lorentz factor is:

$$\gamma(r) = \sqrt{1 + \tilde{\omega}_0{}^2 \tilde{r}^2}. \qquad (12.18)$$

The constant of proportionality between the radial coordinate and the orthogonal angular coordinate is 2π. Therefore:

$$\tilde{\omega}_0 = \frac{\tilde{v}_{max}}{2\pi}. \qquad (12.19)$$

Further than a distance of $1/\tilde{\omega}_0$ from the center, the tangential velocity approaches a constant velocity, $\tilde{v}_{max}$. From this distance outward, on the plane of a revolving coordinate system, the equation of a radial spiral geodesic as measured in K' and transformed onto the coordinate system of a non-revolving observer is:

$$\tilde{r} = \frac{\theta}{\tilde{\omega}_0}. \qquad (12.20)$$

Equation Describing the Flat Velocity Profile

Applying equation (12.12) to the measured tangential velocity using the shifting of spectral lines, v_{tan}, by a non-revolving observer, we now have:

$$v_{tan} = v_{max} \cdot \frac{\tilde{\omega}_0 \tilde{r}}{\sqrt{1 + \tilde{\omega}_0{}^2 \tilde{r}^2}}. \qquad (12.21)$$

As $\tilde{r}$ increases, v_{tan} approaches an asymptote for the maximum tangential velocity as determined by the effects of the Lorentz transformation, which we denoted as $\tilde{v}_{max}$, as a unit-less number, and as v_{max}, in equation (12.21). This completes the derivation of two important functions. One is equation (12.20), which is the equation of a spiral describing a geodesic in K', and the other is equation (12.21), which describes the *apparent* tangential velocity of coordinates in K' calculated by measuring the shifting of spectral lines. We now apply these equations to observations of galaxies. Note that the value of v_{tan} is a measure of velocity using spectral lines. Because of Relativistic effects, v_{tan} has a different value than the tangential velocity which would be given by measuring arc length traversed per unit time. Therefore $v(r)$ of equation (12.7) is not the same as v_{tan}.

12.5.4   Flat Velocity Rotation Profiles

The flat velocity rotation curve of galaxies indicates that nearly all the stars within a galaxy appear to have the same tangential velocity. The model presented here shows this is caused by a discrepancy between the measure of tangential velocity using shifting of spectral lines and actual tangential velocity: that is, distance travelled divided by elapsed time. As a result, the measure of tangential velocities are well below relativistic speeds. In effect, the shifting of spectral lines does not measure the actual tangential velocity of stars within the galaxy.

An Examination of the Rotation Curve of NGC 3198

The rotation curve of NGC 3198 is shown in Figure 12.10. We have curve-fitted equation (12.21) to the observations of Begeman and overlaid the result atop the observed values. The data points provided by Begeman have a reported error in rotational velocity of 5 km/s and an error in angular measure of 15″ of arc. The calculated fit, shown as a continuous line, has a normalized sigma of 0.04 from the data points provided and yields a fitted v_{max} of 152.9 km/s $\pm$ 3.78 km/s. We see that there is a significant correlation between observed values of the rotation curve and the presented model.

There are two parameters involved in obtaining the calculated fit. One is v_{max} and the other is an angular-distance ratio to couple the values given by Begeman in arcmin and the values used in equation (12.21) in ly. A linear regression was used where v^2/R^2_{arcmin} was the ordinate and v^2 the abscissa. v is the measured tangential velocity in km/s and R_{arcmin} is the angular distance from the center of the galaxy in arcmin.

This ratio can also be used to estimate the distance to the galaxy although there is a fairly significant degree of allowable error. Fitting the data presented by Begeman, a distance of 19.2 MPc was found. However, there are discrepancies and difficulties in accurate measurements near the center of the galaxy. Uncertainties in radial measures are at about 15 arcsec as reported by Begeman. The beam width is 30 seconds of arc and the CLEAN software can remove about 1/3 of beam smearing. If the values of radial distance have a discrepancy of 10 seconds of arc, there is a variation in the measure of distance from 12.7 to 48.4 MPc. While (12.21) can give some indication of distance, the problems with finding the center of rotation, coupled with error allowances, near the center of the galaxy, involves a high degree of error. However, future work on improving methods of finding a more accurate distance modulus from rotation profiles may prove very fruitful.

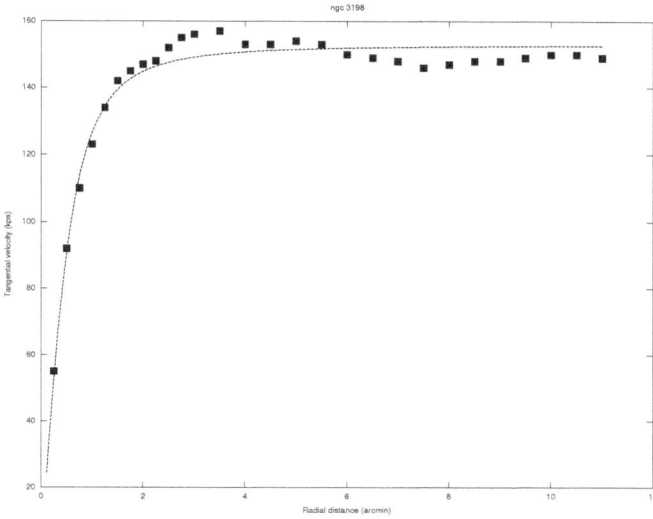

Figure 12.10: Begeman's rotational data overlaid with a curve fit of equation (12.21). The fit yields a normalized sigma of 0.04 and a v_{max} of 152.9 km/s.

Rotation Curves of 878 Galaxies

We have also applied equation (12.21) to 878 velocity rotation profiles from Salucci, [Persic & Salucci (1995)], to obtain an average normalized standard error of 0.0756 with a standard deviation of .049. (Some examples are shown in Figure 12.11).

Table 12.3, at the end of this paper, lists these galaxies with the fitted v_{max} for each galaxy and the normalized σ of the fit of each galaxy's rotation curve to equation (12.21).

12.5.5   Spiral Morphology

The analytical model presented is simply a mathematical spiral as per Equation (12.20) with ω_0 being the constant angular velocity of the galaxy. Digital pho-

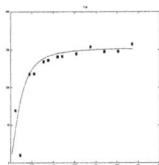 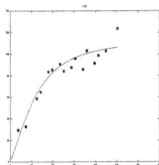 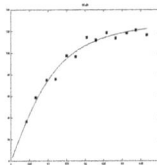

(a) 7-g2   (b) 8-g1   (c) 25-g16

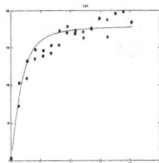 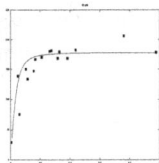 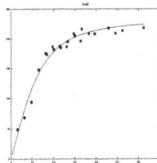

(d) 9-g10   (e) 33-g32   (f) 34-g12

Figure 12.11: Six velocity rotation profiles with Equation (12.21) overlaid. Rotation profiles courtesy of Salucci

tographs from the MAST Digital Sky Survey with maximum response wavelengths between 6400 and 6700 Å, [URL-4], are used in further analysis. Note that the ratio of pixels to arcmin in photographs used is 1.008.

Spiral Morphology of NGC 3198

A v_{max} of 151 km/s for NGC 3198, as given by Begeman, was substituted into Equations (12.19) - (12.20). The resultant curve as described by a geodesic traced out on an equivalent rotating coordinate system to a photograph of the galaxy as in Figure 12.12 (b) was then drawn. This curve is shown in Figure 12.12 (a). It is a graph of a double Archimedes' spiral which closely resembles the photo of NGC 3198. Note the scale of the graph is in thousands of ly as per equation (12.20). There appears to be a remarkable morphological similarity and a possibility of determining the intrinsic size of the galaxy itself.

Measuring the Spiral Pitch of NGC 4321

In the above description of a spiral transformation it can be seen that the resultant spiral shape of material adhering to general relativity in a rotating coordinate system is the result of the transformation from one reference frame to another while conserving the metric. The spiral described by Equation (12.20) is valid for the region where the tangential velocity of material appears to be a constant with respect to radial distance from the center of a rotating coordinate system. This becomes valid when $r >> 1/\omega_0$. (Note, r is in units of ly and ω_0 is in units of radians per year). Thus the outer regions of a galaxy, where a constant rotation profile is well established, can be expected to manifest a constant pitch.

Figure 12.13 is a photo of NGC 4321 with a spiral overlaid according to Equation (12.20) using a value of -32.15 arcsec per radian as a value of κ. This approximates the well-defined spiral portion of NGC 4321. The resultant spiral of this pitch would have an arm spacing close to 32π, or about 100 pixels. Let us define arm spacing as the radial distance between distinct local maxima in the luminosity profile on a line taken through the center of the galaxy, viewing the galaxy as from above. This can be very accurately measured using an FFT of luminosity along this line. If a galaxy is inclined on the celestial sphere, then this line would be oriented along the major axis of the galaxy for the measure of arm spacing to be valid. However, NGC 4321 is seen as a spiral galaxy from almost directly above and the arm spacing, as defined, would be very close to a constant for all cross-sections.

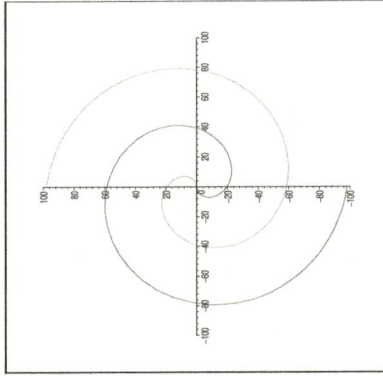

(a) A spiral generated from tracing the outward path of a geodesic upon a rotating polar coordinate system. The scale is in thousands of light years.

(b) NGC 3198 in Ursa Major. NGC 3198 is classified SBc(R), [Youman].

Figure 12.12: A double-arm Archimedes' Spiral is shown in Figure 10(a) and a photograph of NGC 3198 is in Figure 10(b). There appears to be a morphological similarity between the two structures which suggests that an analytical model based on a spiral shape is possible in order to describe galactic morphology and parameters.

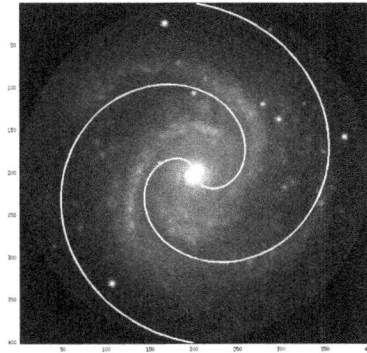

Figure 12.13: NGC 4321 with spiral overlays according to Equation (12.20) with a pitch of -32 arcsec per radian. Note the outer portions of the spiral fall along the path of greater luminosity in the photo of the galaxy.

This was investigated by taking 360 luminosity cross sections through the center of the galaxy at half-degree intervals, applying an FFT along these cross sections, and examining the results to see if a value very close to 100 arcsec would appear. A consistent value of 100.99 arcsec presents itself with values of 134.42 and 80.81 arcsec above and below. This analysis is graphically shown in Figures 12.14 to 12.17.

Measuring the Distance to NGC 3198 using Spiral Pitch

In order to apply a model of spiral galaxies as derived in Equation (12.20), we only require a measure of v_{max}. The resultant spiral would give us the absolute size of any spiral galaxy. If the distance to the galaxy is known, we can determine the scale. The scale then becomes a distance modulus for galaxies. This distance modulus can be determined by the relationship between v_{max} and κ. The scale can be determined, independently of distance, by comparing a derived value of κ in units of ly per radian from v_{max}, and the observed value of κ from digital photographs. This makes the measure of κ the critical parameter in the application of the model to determine the galaxy's distance.

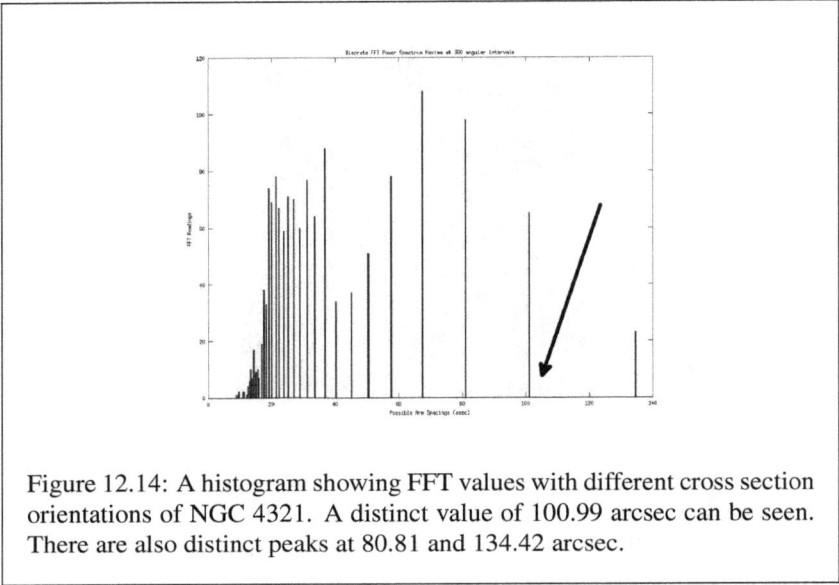

Figure 12.14: A histogram showing FFT values with different cross section orientations of NGC 4321. A distinct value of 100.99 arcsec can be seen. There are also distinct peaks at 80.81 and 134.42 arcsec.

Figure 12.18 is a photo of NGC 3198 in which the pixels have been transformed to display the galaxy as seen from directly above. The figure was overlaid with a spiral according to Equation (12.20) using a value of -26 arcsec per radian as a value of κ. This approximates the well-defined spiral portion of NGC 3198. The resultant spiral of this pitch would have an arm spacing of 26π, or about 82 arcsec.

This property was again investigated by taking 360 luminosity cross sections through the center of the galaxy at half-degree intervals, applying an FFT along these cross sections, and examining the results to see if a value very close to 82 arcsec would appear. A consistent value of 80.83 arcsec presents itself with values of 100.77 and 67.12 arcsec above and below. This analysis is graphically shown in Figures 12.19 to 12.22.

We can validate our model by comparing the predicted intrinsic size of the spiral to the apparent size of a galaxy to estimate its distance. We then compare the derived distance measure to other distance measures in order to determine a degree of validation for the model. The derived distance modulus is given by the equation:

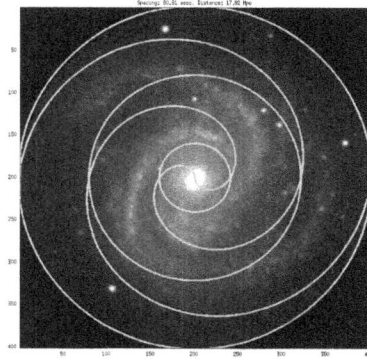

Figure 12.15: Digital photograph of NGC 4321 with circles overlaid at 80.81 arcsec intervals. The intervals indicate a spacing somewhat smaller than the pitch of the galaxy. A spiral overlay having a corresponding pitch is also overlaid. It can be seen that the overlaid spiral is significantly more "wound up" than the spiral shown by the brighter regions of the galaxy itself.

$$D = \frac{3.12 \times 10^9}{v_{max} \times \alpha_s} \tag{12.22}$$

where 3.12×10^9 is in pc arcmin km/s.

In the case of NGC 3198, v_{max} is taken as 152.9 km/s based on the above described curve fit of the velocity profile and α_s is an angular measure of spiral pitch equal to 80.83 arcsec, or 1.347 arcmin. From equation 12.22 we determine NGC 3198 to be 15.15 (± 2.5) Mpc distant. The allowable error in v_{max} is calculated as 7.5 km/s and in α_s as 0.14 arcmin.

These measurements compare to 13.8 Mpc by Freedman, [Freedman (2001)], 12 Mpc using Cepheid variables and 13.8 Mpc using Tully-Fisher, [Tully et al (2008)], 10.92 Mpc using redshift by Crook, [Crook et al (2007)], 14.5 using Cepheid variables by Ferrarese, [Ferrarese et al (2000)] and 17 Mpc by Gil de Pas, [Gil De Pas et al (2007)].

These measures have a mean of 13.9 Mpc with a standard deviation of 2.0 Mpc of the seven measurements presented here.

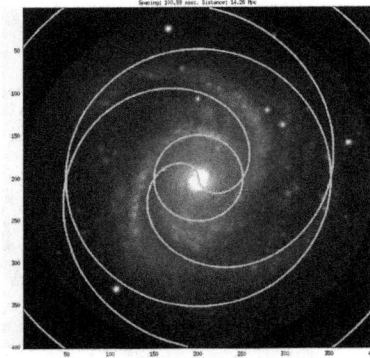

Figure 12.16: Digital photograph of NGC 4321 with circles overlaid at 100.99 arcsec intervals. The intervals indicate a spacing which matches the pitch of the galaxy. A spiral overlay having a corresponding pitch is also overlaid. It can be seen that the overlaid spiral is as "wound up" as the spiral shown by the brighter regions of the galaxy itself.

12.5.6 Using Distance Measures to Validate the Model

Distance measures to galaxies can be used to determine a degree of validation for the presented model. There are three different distance measures that will be presented here and compared to predictions. We shall use distance measures using Cepheid variables, the Tully-Fisher relationship and behavior of water masers.

Comparing Distance Measures using Cepheid Variables

We have reviewed distance measurements to galaxies made by Ferrarese, [Ferrarese et al (2000)], using Cepheid variables, [Leavitt (1908)], and compared these measurements to measurements made using equation (12.22) and rotation curves, which are cited in Table 12.4 shown at the end of this paper. Figure 12.23 is a presentation of a comparison between Cepheid measurements and equation (12.22) showing a discrepancy from matching a one to one linear fit of 0.016. The confidence variable is 0.9104. Table 12.4 lists the name of the galaxy in the first column. The second column lists estimates of v_{max} from rota-

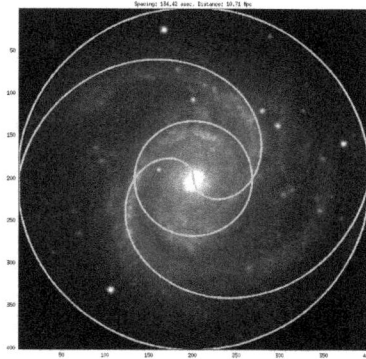

Figure 12.17: Digital photograph of NGC 4321 with circles overlaid at 134.42 arcsec intervals. The intervals indicate a spacing somewhat larger than the pitch of the galaxy. A spiral overlay having a corresponding pitch is also overlaid. It can be seen that the overlaid spiral is significantly less "wound up" than the spiral shown by the brighter regions of the galaxy itself.

tion curves given by the cited papers below the table with an allowance of 10% as listed in the ninth column. The third column lists the measure of α_s from an FFT. The fourth column lists the distance calculated by equation (12.22) and the normalized error in the distance measure is listed in the fifth column. The sixth column lists the magnitude difference from observing Cepheid variables within the galaxy. The seventh column lists the distance measure calculated from using Cepheid variables and the eigth column lists the normalized error in the distance measure using Cepheids. The final column gives references.

Comparing Distance Measures using Water Masers

Another method to measure the distance to galaxies can be found through the behaviour of water masers by Herrnstein, [Herrnstein et al (1999)]. In this method, the magnitudes of orbits of gases containing masers can be measured directly and then compared to the angular measure of these orbits. Herrnstein has measured a distance of 7.3 ± 0.3 Mpc to NGC 4258 using the behavior of water masers within the galaxy while equation (12.22), yields a distance mea-

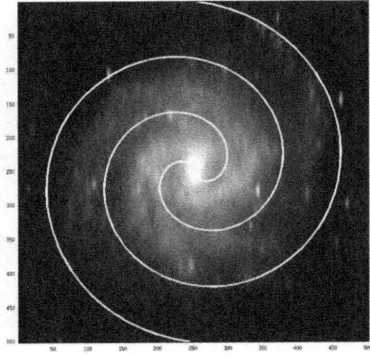

Figure 12.18: NGC 3198 with spiral overlays according to Equation (12.20) with a pitch of -26 arcsec per radian.

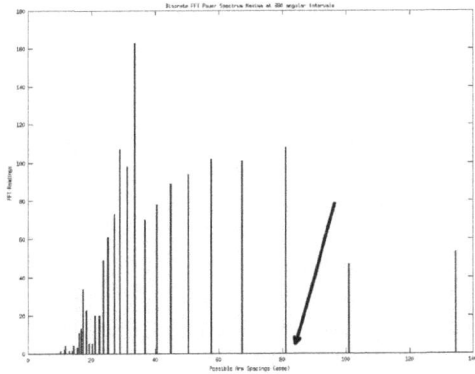

Figure 12.19: A histogram showing FFT values with different cross section orientations of NGC 3198. A distinct value of 80.83 arcsec can be seen. There are also distinct peaks at 67.12 and 100.77 arcsec to either side.

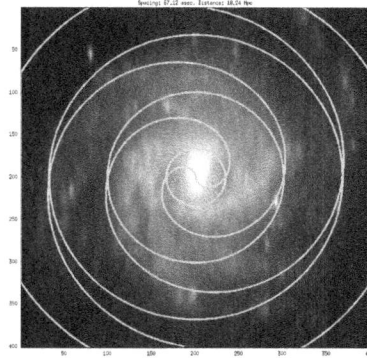

Figure 12.20: Digital photograph of NGC 3198 with circles overlaid at 67.12 arcsec intervals. The intervals indicate a spacing somewhat smaller than the pitch of the galaxy. A spiral overlay having a corresponding pitch is also overlaid. It can be seen that the overlaid spiral is significantly more "wound up" than the spiral shown by the brighter regions of the galaxy itself.

sure of 7.1 ± 0.55 Mpc. Table 12.1 contains rotational velocity measurements by Burbidge, [Burbidge et al (1963)], which reports that NGC 4258 has dusty regions in which there are no emission patches strong enough to be recorded: "The measures between $180''$ and $220''$ on the north west side come from the spiral arm crossed by the spectrograph slit". A spiral pitch of 1.81 ± 0.008 arcmin and v_{max} of 244.0 ± 17.86 km/s was used to determine this measure.

Table 12.1: Rotation velocities of NGC 4258 from Burbidge

Distance from center of galaxy (arcsec)	Tangential velocity (km/s)
185	225
188.6	240
(Continued...)	

Data from Burbidge on rotation velocities of NGC 4258. (Continued)	
Distance from center of galaxy (arcsec)	Tangential velocity (km/s)
192	255
195	255
199.5	255
203.1	255
206.7	240
210.4	270
214	255
217.6	210
221.2	225

Data from Burbidge on rotation velocities of NGC 4258 taken from $185''$ to $220''$ from the center of the galaxy.
The data comes from an area of the galaxy where a galactic arm crosses with the major axis. The average is 244.09 km/s with a standard deviation of 17.86 km/s.

Another distance measurement to a galaxy using water maser behaviour was conducted by Braatz, [Braatz et al (1955)]. Braatz measured the distance to UGC 3789 as 49.9 ± 7.0 Mpc. Unfortunately, no rotation curve for UGC 3789 has been reported for this galaxy. Nevertheless, an H-1 line width is available through NED[1], [Theureau et al (1998)], (see Figure 12.24). Using the spectrum reported for UGC 3789 and reported measurements of the angle of incline of the galaxy, $44.8°$, a v_{max} of 314.33 ± 50.7 km/s was calculated. An FFT across the galaxy's major axis gave a pitch for the galaxy of 11.99 ± 0.7 arcsec. The resultant distance to UGC 3789 using equation (12.22) is 49.7 ± 8 Mpc.

---

[1]This research has made use of the NASA/IPAC Extragalactic Database (NED) which is operated by the Jet Propulsion Laboratory, California Institute of Technology, under contract with the National Aeronautics and Space Administration.

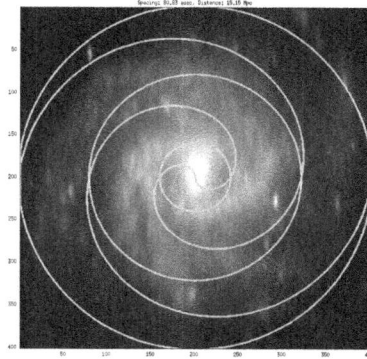

Figure 12.21: Digital photograph of NGC 3198 with circles overlaid at 80.83 arcsec intervals. The intervals indicate a spacing which matches the pitch of the galaxy. A spiral overlay having a corresponding pitch is also overlaid.

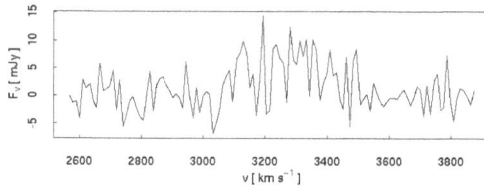

Figure 12.24: Spectral response of signals traversing the major axis of UGC 3789. The line width is not clear and an estimate of 221.145 km/s from 3063.54 km/s to 3505 km/s is submitted. With an angle of incline of $44.8°$ yields an estimated v_{max} of 314.33 km/s.

The measure using equation (12.22) was particularly difficult to make due to the lack of distinguishing spiral shape of the galaxy. The galaxy is fairly distant and it is tightly wound. Note that its large rotation velocity would result

95

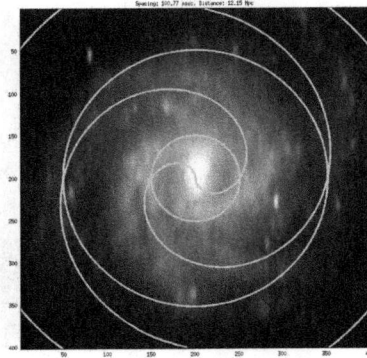

Figure 12.22: Digital photograph of NGC 3198 with circles overlaid at 100.77 arcsec intervals. The intervals indicate a spacing somewhat larger than the pitch of the galaxy. A spiral overlay having a corresponding pitch is also overlaid. It can be seen that the overlaid spiral is significantly less "wound up" than the spiral shown by the brighter regions of the galaxy itself.

in a tightly wound galaxy in accordance with the model presented here. Furthermore, the spectral line width, as in Figure 12.24, is a little indistinct, and six different measures of the b/a ratio are reported in NED. Therefore, the error allowance is quite large. The average b/a ratio was calculated to be 0.71 ± 0.059.

Comparing Distance Measures using Tully-Fisher

Yet another method for measuring the distances to galaxies is the well known Tully-Fisher (T-F), [Tully & Fisher (1977)]. This method involves an observed relationship between the width of spectral lines and luminosity of spiral galaxies. The spectral line widths are caused by the rotation of the galaxy. This rotational velocity is denoted as v_{rot} and corresponds to v_{max}. In Figure 12.25, we present a graph of equation (12.22) measurements vs. distance measurements using T-F. The graph shows a linear fit through the origin with a discrepancy from a one to one match of .0272 and a sigma of 1.38 Mpc. The associated data can be found in Table 12.2. The first column in this table is the name of the galaxy being measured. The second column is the value of v_{max}. The

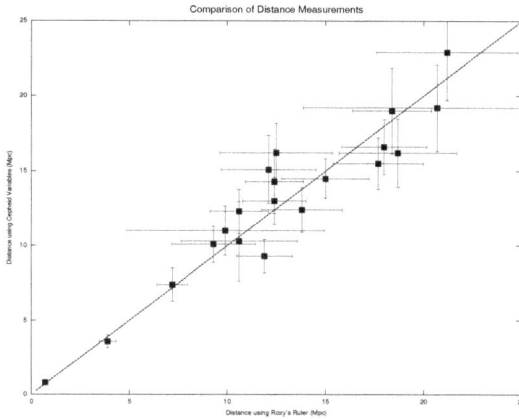

Figure 12.23: On the "y" axis is plotted the Cepheid distance of each galaxy vs. the distance determined by equation (12.22) on the "x" axis. The graphs would have a slope of one for a perfect match. The graph shows a discrepancy from matching a one to one linear fit by 0.016, or 1.6%. The confidence variable is 0.9104. A perfect fit would have a confidence variable of one.

third column is distance measure using T-F as reported in the Simbad database, [URL-5]. The fourth column is the reported error in the T-F measure in Mpc. The fifth column is the distance measure using equation (12.22) and the sixth column is the allowable error of the measure in Mpc. The error bars in Figure 12.25 reflect the reported errors in Table 12.2.

Table 12.2: Table of Tully-Fisher Distances

Name	v_{max}	T-F Distance	Error	Equation (12.22) Distance	Error
	(km/s)	(Mpc)	(Mpc)	(Mpc)	(Mpc)
ESO 284-24	135.073	29	6	25	1.93
ESO 378-11	131.681	50	9	52	5.16
ESO 576-11	146.63	31	6	24	1.40
IC 5078	119.393	19	4	21	0.88
UGCA 17	109.892	23	5	18	1.12
NGC 1090	180.845	27	5	29	0.95
NGC 1163	160.503	39	7	37	2.30

97

Name	v_{max}	T-F Distance	Error	Equation (12.22) Distance	Error
	(km/s)	(Mpc)	(Mpc)	(Mpc)	(Mpc)
NGC 1337	112.9	11	2	16	0.92
NGC 1832	198.743	25	5	27	1.54
NGC 2763	145.929	24	5	23	1.39
NGC 3321	144.019	33	7	33	4.05
NGC 4348	182.247	30	6	29	2.15
NGC 701	125.347	19	4	30	4.33
NGC 7218	128.406	21	4	25	1.55
NGC 7339	156.265	22	4	22	0.95
NGC 755	133.339	19	4	36	2.37
NGC 7606	273.5	32	6	33	13.45

Comparisons of distance measures using equation (12.22) and Tully-Fisher. Data taken from SIMBAD database operated at CDS, Strasbourg, France

12.6   A Quasi-linear Physical Model

If we project an elongated collection of stars onto a single dimension, we can map a four dimensional Minkowski space into L-1 space, [Minkowski (1910)].

Mass and Linear Density

In the model we are presenting, gravitationally self bound particles appear to be oriented along the path of the spiral shaped geodesic as in equation (12.20). A mapping of material in K' into L-1 space through rotation onto a linear axis is straightforward. Using the Lebesgue, [Lebesgue (1902)] measure of linear density in a singular dimension, we have:

$$\rho_l = \frac{v_{max}^2}{2G}. \tag{12.23}$$

Using previously described distance measures and the angular length of the major axis, the intrinsic major axis length, L, can be determined. Using equation (12.26) we can determine the mass of the galaxy as:

$$M_g = L\rho_l. \tag{12.24}$$

From this we can calculate the angular momentum of a galaxy as:

$$l = v_{max}\rho_l L^2 \tag{12.25}$$

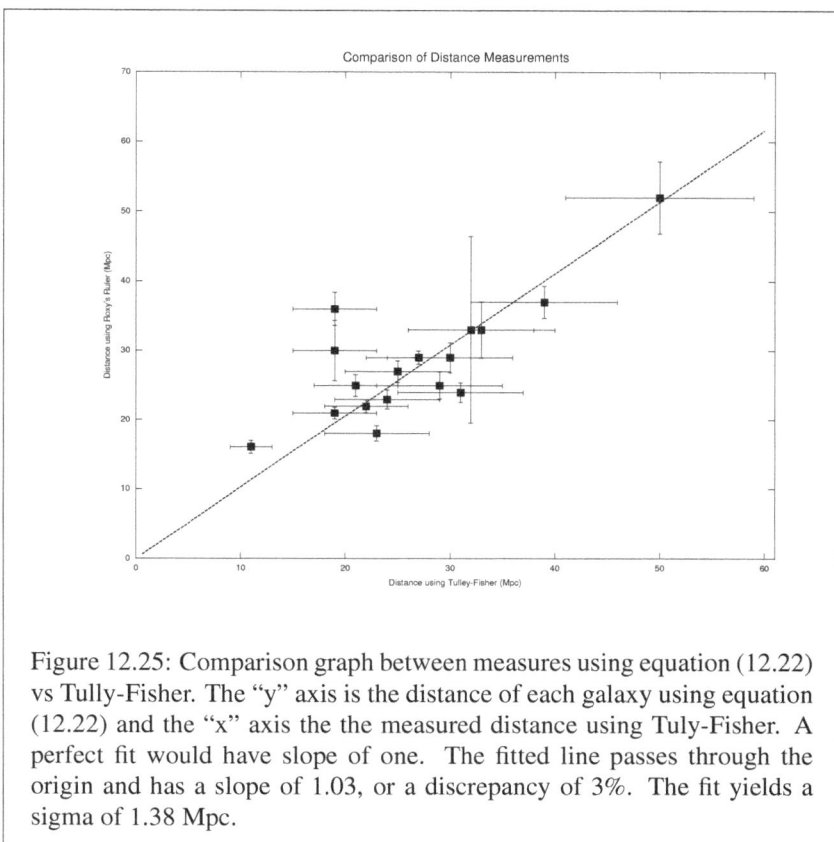

Figure 12.25: Comparison graph between measures using equation (12.22) vs Tully-Fisher. The "y" axis is the distance of each galaxy using equation (12.22) and the "x" axis the the measured distance using Tuly-Fisher. A perfect fit would have slope of one. The fitted line passes through the origin and has a slope of 1.03, or a discrepancy of 3%. The fit yields a sigma of 1.38 Mpc.

where ρ_l is linear density.

We derive the overall linear density of a galaxy to be within an order of magnitude of 10^{20} kg per meter using Equation (12.26). The resultant body of particles, in a linear orientation, is highly viscous. Using this linear density, a galaxy having a diameter of 300 million ly would have a mass within an order of magnitude of 10^{11} solar masses.

12.6.1 CONCLUSIONS

Our calculations take into account Minkowski, Lorentz and Einstein's principles on time and space dilation. We have come to the following conclusions:

Masking Tangential Velocities

Considerations of curvature, relativity, effects of motion, etc. must be thought of locally. Currently, the circular orbital speed of stars within distant galaxies is determined by measuring the shifting of spectral lines of sections of the galaxy. It is not determined by the distance travelled by stars divided by the time it takes for them to get farther along in their orbit. The distances the stars are travelling are vast and we have not observed galaxies for a long enough time to detect circular motion directly. Shifts in spectral lines are not restricted to only the effects of distance travelled divided by time.

In a star's local reference frame, its clocks and rulers differ from the clocks and rulers of other stars in the galaxy due to the variation of the Lorentz transformations at different radial distances from the center. The clocks of each Hydrogen atom in a distant galaxy vary from the clocks of local Hydrogen atoms since the distant orbiting atoms are in a region of curved space and time. Therefore, the shifting of spectral lines of Hydrogen atoms in a distant rotating galaxy is affected by both the Lorentz transformation of Doppler effects and from the Lorentz transformation as a result of the curvature of space-time in a rotating system. Both Lorentz transformations, one varying directly with r and the other varying inversely, cancel in outer regions of the galaxy and result in the observed constant velocity rotation profile.

We furthermore conclude:

1. General relativity explains the flat velocity rotation profile and morphology of spiral galaxies.

2. Spiral galaxies are gravitationally self bound.

3. Galaxies are gravitationally viscous.

4. Galaxies are morphologically stable.

5. Einstein's general theory of relativity and Newton's principles of gravitational attraction hold over very great distances.

Acknowledgments

The authors wish to acknowledge the kindness of P. Salucci for access to his data following the loss of the original data tape of Mathewson's observations as a result of a library fire in Australia, the technical assistance of W. Israel, the wide band support of SARA and the encouragement, suggestions and guidance of Roxy.

12.6.2 Tables of Data

Table 12.3: Table of galaxies from Persic & Salucci.

name	v_{max} (ks^{-1})	σ (norm)	name	v_{max} (ks^{-1})	σ (norm)	name	v_{max} (ks^{-1})	σ (norm)
101-g20	198.9	0.08	120-g16	145.5	0.09	1547-02	209.2	0.03
101-g5	200.4	0.07	123-g15	217.9	0.07	156-g36	151.5	0.36
102-g10	180.7	0.04	123-g16	245.9	0.08	157-g20	143.5	0.03
102-g15	180.3	0.05	123-g23	157.5	0.04	157-g38	91.0	0.12
102-g7	245.2	0.06	123-g9	146.3	0.08	160-g2	217.0	0.15
103-g13	223.2	0.07	124-g15	128.9	0.06	162-g15	86.7	0.11
103-g15	154.6	0.11	13-g16	131.9	0.03	162-g17	65.8	0.03
103-g39	73.6	0.13	13-g18	148.5	0.06	163-g11	168.8	0.04
104-g52	108.3	0.06	140-g24	225.4	0.06	163-g14	180.9	0.09
105-g20	230.6	0.07	140-g25	91.8	0.04	181-g2	296.7	0.12
105-g3	165.7	0.06	140-g28	111.8	0.16	183-g14	141.0	0.07
106-g12	139.0	0.07	140-g33	220.0	0.51	183-g5	97.7	0.06
106-g8	157.0	0.08	140-g34	136.0	0.15	184-g51	262.1	0.04
107-g24	178.6	0.08	140-g43	146.4	0.12	184-g54	186.0	0.09
107-g36	197.2	0.07	141-g20	254.5	0.04	184-g60	88.3	0.11
108-g13	131.6	0.05	141-g23	114.4	0.13	184-g63	178.7	0.07
108-g6	172.8	0.12	141-g34	287.0	0.04	184-g67	231.4	0.06
109-g32	124.6	0.07	141-g37	312.0	0.09	184-g74	206.6	0.20
10-g4	117.6	0.06	141-g9	236.1	0.09	184-g8	128.0	0.19
111-g9	147.0	0.05	142-g30	203.5	0.06	185-g11	230.5	0.08
112-g10	165.0	0.18	142-g35	261.7	0.06	185-g36	176.0	0.04
113-g21	112.1	0.06	143-g10	37.8	0.05	185-g68	114.1	0.06
113-g6	238.1	0.04	143-g6	124.4	0.34	185-g70	153.7	0.08
114-g21	174.6	0.11	143-g8	81.5	0.15	186-g26	77.6	0.07
116-g12	126.8	0.02	145-g18	174.9	0.13	186-g75	123.5	0.06
116-g14	155.6	0.05	145-g22	200.1	0.07	186-g8	134.3	0.07
117-g18	201.0	0.06	146-g6	143.6	0.04	187-g39	114.2	0.08
117-g19	206.6	0.05	151-g30	212.8	0.05	187-g8	138.0	0.04
189-g12	257.0	0.03	231-g6	99.7	0.06	251-g10	238.1	0.11
18-g13	256.2	0.07	233-g36	117.0	0.07	251-g6	170.6	0.06
18-g15	196.6	0.10	233-g41	286.2	0.10	25-g16	131.7	0.10
196-g11	127.9	0.07	233-g42	103.6	0.06	264-g43	260.7	0.08
197-g24	152.5	0.04	234-g13	146.7	0.08	264-g48	178.8	0.08
197-g2	166.0	0.06	234-g32	170.1	0.19	265-g16	158.2	0.05
1-g7	118.4	0.10	235-g16	251.8	0.05	265-g2	57.3	0.05
200-g3	102.4	0.03	235-g20	154.9	0.04	266-g8	106.4	0.03
202-g26	135.8	0.08	236-g37	180.6	0.07	267-g29	198.9	0.11
202-g35	114.1	0.06	237-g15	129.7	0.28	267-g38	223.7	0.09
204-g19	136.7	0.06	237-g49	101.2	0.07	268-g11	232.8	0.11
205-g2	85.1	0.08	238-g24	210.6	0.04	268-g33	230.3	0.05
206-g17	67.7	0.06	239-g17	106.6	0.12	269-g15	242.8	0.06
208-g31	195.7	0.05	240-g11	219.8	0.02	269-g19	177.3	0.02
215-g39	148.6	0.09	240-g13	189.0	0.07	269-g48	99.4	0.11
216-g21	180.4	0.06	241-g21	246.1	0.09	269-g49	140.2	0.11
216-g8	202.8	0.05	242-g18	115.0	0.11	269-g52	193.6	0.04
219-g14	312.6	0.08	243-g14	162.3	0.04	269-g61	369.9	0.07
21-g3	105.6	0.08	243-g34	349.2	0.33	269-g75	110.9	0.06
220-g8	169.0	0.06	243-g36	178.0	0.07	269-g82	125.5	0.15
221-g21	156.0	0.04	243-g8	209.7	0.08	26-g6	116.3	0.04
221-g22	133.5	0.12	244-g31	257.3	0.13	271-g22	180.8	0.05
22-g12	141.4	0.04	244-g43	158.8	0.12	273-g6	236.5	0.08
22-g3	114.7	0.09	245-g10	174.2	0.06	27-g17	215.1	0.06
231-g11	243.1	0.10	249-g16	186.9	0.02	27-g24	188.8	0.06
231-g23	242.9	0.07	249-g35	72.9	0.06	27-g8	176.4	0.06
231-g25	209.0	0.04	24-g19	217.5	0.11	280-g13	276.9	0.10
231-g29	125.4	0.07	250-g17	281.5	0.14	281-g38	212.4	0.08
282-g35	136.7	0.13	303-g14	288.8	0.11	322-g82	213.8	0.04
282-g3	189.0	0.05	304-g16	204.5	0.11	322-g85	111.4	0.10
284-g13	181.5	0.08	305-g14	141.0	0.08	322-g87	152.1	0.09
284-g21	131.2	0.07	305-g6	159.6	0.02	322-g93	108.9	0.11
284-g24	135.1	0.06	306-g2	106.7	0.05	323-g27	201.7	0.09
284-g29	149.5	0.04	306-g32	172.4	0.04	323-g33	146.0	0.07
284-g39	120.8	0.08	308-g23	161.9	0.11	323-g41	152.1	0.08
285-g27	279.3	0.18	309-g17	261.0	0.09	323-g42	129.5	0.10
285-g40	239.5	0.07	309-g5	87.8	0.11	325-g27	114.0	0.12
286-g16	189.5	0.03	30-g9	329.4	0.03	325-g50	89.5	0.06
286-g18	332.7	0.03	310-g2	234.6	0.07	327-g27	120.7	0.09
287-g13	177.5	0.03	317-g32	240.8	0.10	327-g31	129.4	0.04
289-g10	107.5	0.03	317-g52	191.3	0.06	328-g15	203.4	0.07
290-g22	151.1	0.05	319-g16	95.3	0.05	328-g3	227.7	0.05
290-g35	205.0	0.05	319-g26	117.5	0.09	328-g41	238.4	0.05
291-g10	211.4	0.03	31-g18	177.6	0.06	328-g43	107.8	0.05
291-g24	76.9	0.08	31-g5	203.2	0.04	328-g46	240.7	0.07

(Continued...)

Table of galaxies from Persic & Salucci. (Continued)

name	v_{max} (ks^{-1})	σ (normalized)	name	v_{max} (ks^{-1})	σ (normalized)	name	v_{max} (ks^{-1})	σ (normalized)
296-g26	477.4	0.09	320-g24	131.0	0.05	329-g7	270.7	0.05
297-g37	166.6	0.11	320-g26	230.0	0.05	32-g18	207.0	0.06
298-g16	323.2	0.09	320-g2	369.5	0.20	336-g13	194.0	0.05
298-g29	238.6	0.16	321-g10	136.9	0.08	336-g6	264.8	0.19
298-g36	128.8	0.07	321-g17	137.2	0.04	337-g22	145.4	0.07
298-g8	150.1	0.06	321-g1	179.1	0.10	337-g6	188.3	0.09
299-g18	172.7	0.07	322-g33	46.4	0.09	338-g22	119.0	0.05
299-g4	186.2	0.05	322-g36	152.5	0.10	339-g36	162.2	0.05
2-g12	193.9	0.06	322-g45	126.5	0.07	33-g22	185.9	0.07
302-g7	135.2	0.09	322-g48	126.5	0.07	33-g32	177.8	0.09
302-g9	73.6	0.03	322-g55	175.2	0.16	340-g26	169.3	0.04
342-g43	168.7	0.13	354-g47	232.5	0.08	374-g26	135.3	0.08
343-g18	146.5	0.07	355-g26	121.6	0.05	374-g27	253.2	0.04
343-g28	109.4	0.11	356-g15	229.6	0.07	374-g29	133.5	0.08
344-g20	471.6	0.13	356-g18	66.2	0.08	374-g3	145.2	0.04
346-g14	107.9	0.04	357-g16	79.6	0.05	374-g49	211.3	0.06
346-g1	118.0	0.09	357-g19	126.4	0.06	374-g8	71.3	0.06
346-g26	98.4	0.02	357-g3	147.2	0.05	375-g12	283.8	0.10
347-g28	96.0	0.02	358-g17	264.8	0.06	375-g26	171.9	0.03
347-g33	186.6	0.04	358-g58	165.8	0.07	375-g29	175.4	0.05
347-g34	116.9	0.02	358-g63	113.5	0.08	375-g2	182.3	0.11
349-g32	296.4	0.08	358-g9	101.9	0.08	375-g47	143.5	0.12
349-g33	204.7	0.07	359-g6	76.2	0.04	376-g2	233.8	0.09
34-g12	232.6	0.05	35-g18	129.2	0.06	377-g10	193.0	0.05
350-g23	230.5	0.01	35-g3	89.6	0.05	377-g11	341.3	0.10
351-g18	129.4	0.07	361-g12	136.2	0.06	377-g31	175.3	0.06
351-g1	108.2	0.09	362-g11	129.4	0.01	378-g11	131.7	0.06
351-g28	116.7	0.19	363-g23	174.1	0.05	379-g6	176.4	0.05
352-g14	200.9	0.12	363-g7	84.7	0.06	380-g14	185.3	0.11
352-g15	137.6	0.11	365-g28	187.7	0.06	380-g19	239.0	0.04
352-g24	166.2	0.11	365-g31	199.3	0.18	380-g23	113.1	0.05
352-g27	219.1	0.06	366-g4	142.9	0.13	380-g24	154.3	0.08
352-g50	148.7	0.08	366-g9	107.9	0.14	380-g25	66.0	0.17
352-g53	251.4	0.05	36-g19	209.0	0.03	380-g29	87.5	0.37
353-g14	153.9	0.05	373-g12	82.7	0.09	380-g2	54.6	0.07
353-g26	218.2	0.19	373-g21	105.9	0.07	381-g51	249.6	0.06
353-g2	110.7	0.11	373-g29	140.9	0.06	382-g32	225.6	0.12
354-g17	182.8	0.06	374-g10	140.8	0.04	382-g41	95.1	0.10
354-g46	208.1	0.10	374-g11	201.3	0.05	382-g4	170.5	0.09
382-g58	306.7	0.13	406-g26	116.1	0.03	422-g12	340.2	0.10
383-g2	209.3	0.11	406-g33	133.0	0.03	422-g23	248.5	0.19
383-g55	269.1	0.10	40-g12	199.0	0.04	426-g8	176.3	0.04
383-g67	119.4	0.05	410-g19	187.8	0.04	427-g14	97.7	0.09
383-g88	191.0	0.12	410-g27	164.9	0.10	427-g2	210.9	0.03
385-g12	187.5	0.15	411-g3	215.3	0.10	42-g10	173.7	0.05
385-g8	149.8	0.04	412-g15	150.1	0.06	42-g3	239.4	0.04
386-g43	308.8	0.07	412-g21	201.3	0.06	433-g10	163.1	0.09
386-g44	176.3	0.13	413-g14	262.9	0.12	433-g15	126.7	0.05
386-g6	160.4	0.08	413-g23	143.3	0.07	433-g17	365.0	0.11
387-g20	179.2	0.09	414-g25	190.3	0.06	434-g23	145.6	0.07
387-g26	229.4	0.05	414-g8	96.6	0.16	435-g10	138.0	0.05
387-g4	246.6	0.07	415-g10	112.4	0.06	435-g14	170.4	0.08
38-g12	152.9	0.06	415-g15	210.1	0.07	435-g19	101.6	0.09
398-g20	210.0	0.09	415-g28	130.3	0.10	435-g24	183.1	0.07
399-g23	220.4	0.07	416-g37	208.3	0.06	435-g34	135.0	0.05
3-g3	248.7	0.08	416-g41	202.2	0.12	435-g50	124.7	0.06
3-g4	192.2	0.09	417-g18	166.5	0.07	435-g51	130.5	0.09
400-g21	128.8	0.04	418-g15	139.3	0.12	435-g5	326.2	0.08
400-g37	122.5	0.06	418-g1	115.8	0.04	436-g34	271.8	0.07
400-g5	182.8	0.06	418-g8	77.7	0.06	436-g39	215.8	0.06
401-g3	241.5	0.09	418-g9	90.7	0.07	436-g3	162.8	0.03
403-g16	200.4	0.12	419-g3	144.7	0.05	437-g18	126.1	0.12
403-g31	95.7	0.09	419-g4	180.7	0.03	437-g22	149.6	0.07
404-g18	66.7	0.09	41-g6	92.9	0.16	437-g25	159.4	0.06
404-g31	130.3	0.08	41-g9	193.5	0.03	437-g31	136.8	0.14
404-g45	141.2	0.05	420-g3	168.4	0.05	437-g35	103.1	0.10
405-g5	224.1	0.20	422-g10	284.4	0.09	437-g47	78.9	0.07
437-g54	150.0	0.05	444-g86	222.6	0.07	461-g3	243.9	0.08
437-g56	193.3	0.40	445-g19	226.8	0.07	461-g44	186.1	0.09
437-g69	87.8	0.17	445-g26	202.2	0.22	462-g16	118.3	0.06
437-g71	54.8	0.13	445-g39	297.3	0.11	463-g25	243.8	0.04
437-g72	225.9	0.10	446-g18	241.9	0.04	466-g13	204.0	0.09
438-g15	158.5	0.06	446-g1	196.1	0.09	466-g27	229.7	0.13
438-g18	179.3	0.06	446-g23	273.5	0.04	466-g28	165.5	0.07
439-g11	72.3	0.06	446-g2	217.3	0.09	466-g5	133.6	0.07
439-g18	446.8	0.07	446-g44	134.8	0.03	467-g11	302.7	0.06
439-g20	225.7	0.09	446-g51	148.9	0.05	467-g23	223.0	0.07
439-g9	301.5	0.08	446-g53	50.8	0.05	467-g27	196.1	0.06
43-g8	315.1	0.08	446-g58	223.3	0.12	467-g36	200.6	0.08

(Continued...)

102

name	v_{max} $(k\,s^{-1})$	σ (normalized)	name	v_{max} $(k\,s^{-1})$	σ (normalized)	name	v_{max} $(k\,s^{-1})$	σ (normalized)
440-g51	101.8	0.10	447-g19	252.4	0.15	467-g51	92.2	0.04
441-g11	61.5	0.06	447-g21	211.1	0.08	468-g11	178.1	0.05
441-g24	111.4	0.06	447-g23	163.5	0.06	468-g23	95.0	0.02
441-g2	123.5	0.03	448-g13	276.8	0.04	469-g22	186.0	0.22
442-g24	129.4	0.06	44-g13	365.5	0.07	46-g8	99.9	0.08
442-g2	74.0	0.06	44-g1	151.3	0.08	471-g2	240.5	0.14
443-g38	253.2	0.10	450-g18	108.8	0.08	472-g10	146.6	0.08
443-g42	282.4	0.06	452-g8	133.8	0.05	474-g19	126.2	0.17
443-g59	100.7	0.06	459-g14	129.2	0.05	474-g39	179.2	0.06
443-g80	130.2	0.08	459-g6	242.4	0.07	474-g5	133.7	0.04
444-g10	176.4	0.05	460-g25	265.5	0.15	476-g15	150.6	0.02
444-g14	132.1	0.08	460-g29	391.7	0.07	476-g16	211.8	0.08
444-g1	251.3	0.10	460-g31	237.0	0.05	476-g25	188.5	0.09
444-g21	101.9	0.06	460-g8	172.8	0.06	476-g5	266.1	0.08
444-g33	61.0	0.09	461-g10	156.5	0.07	477-g16	108.7	0.05
444-g47	162.1	0.07	461-g25	162.7	0.06	477-g18	196.8	0.06
478-g11	120.9	0.04	490-g36	115.8	0.06	507-g2	159.9	0.09
479-g1	121.9	0.07	490-g45	100.2	0.04	507-g56	210.0	0.06
47-g10	211.3	0.05	496-g19	131.2	0.09	507-g62	164.6	0.07
481-g11	150.4	0.07	497-g14	295.1	0.15	507-g7	299.0	0.04
481-g13	176.1	0.02	497-g18	242.0	0.05	508-g11	112.4	0.03
481-g2	148.7	0.02	497-g34	205.2	0.06	508-g60	152.0	0.06
482-g1	158.5	0.08	498-g3	165.5	0.04	509-g35	211.1	0.09
482-g2	184.8	0.15	499-g22	114.7	0.05	509-g44	258.1	0.10
482-g35	132.0	0.05	499-g26	135.0	0.05	509-g45	130.3	0.04
482-g41	189.6	0.06	499-g39	187.3	0.04	509-g74	158.3	0.04
482-g43	161.8	0.09	499-g4	144.4	0.12	509-g80	260.2	0.21
482-g46	92.2	0.03	499-g5	158.9	0.04	509-g91	136.3	0.04
483-g12	167.6	0.11	4-g19	139.0	0.08	510-g40	134.1	0.05
483-g2	112.8	0.07	501-g11	129.5	0.05	511-g46	120.6	0.07
483-g6	175.4	0.04	501-g1	156.5	0.09	512-g12	186.7	0.07
484-g25	157.7	0.17	501-g68	154.5	0.10	514-g10	161.8	0.07
485-g12	152.7	0.05	501-g69	92.1	0.05	51-g18	96.5	0.05
485-g4	145.2	0.04	501-g75	167.5	0.04	526-g11	134.5	0.08
487-g19	99.1	0.06	501-g80	71.8	0.04	527-g11	220.7	0.09
487-g2	177.8	0.04	501-g86	172.9	0.15	527-g19	219.5	0.09
488-g44	117.6	0.08	501-g97	267.9	0.08	527-g21	132.5	0.10
488-g54	180.8	0.04	502-g12	151.0	0.07	528-g17	140.3	0.09
489-g11	140.1	0.06	502-g13	132.7	0.01	528-g34	165.3	0.10
489-g6	117.1	0.05	502-g2	206.0	0.07	530-g34	223.8	0.08
48-g8	224.4	0.03	505-g8	81.1	0.16	531-g22	182.2	0.03
490-g10	135.7	0.06	506-g2	234.6	0.04	531-g25	177.2	0.05
490-g14	116.6	0.05	506-g4	355.0	0.07	532-g14	62.5	0.06
490-g28	55.2	0.07	507-g11	213.6	0.09	533-g48	151.3	0.05
533-g4	170.1	0.04	547-g14	243.4	0.05	554-g28	118.2	0.09
533-g53	163.6	0.01	547-g1	96.2	0.07	554-g29	129.8	0.05
533-g8	175.8	0.08	547-g24	120.0	0.05	554-g34	177.3	0.04
534-g24	157.3	0.08	547-g31	160.0	0.06	555-g16	258.1	0.04
534-g31	282.6	0.07	547-g32	195.4	0.08	555-g22	100.2	0.03
534-g3	167.3	0.15	547-g4	124.9	0.13	555-g29	132.9	0.07
534-g9	226.4	0.06	548-g21	66.3	0.07	555-g2	143.9	0.07
535-g15	212.9	0.08	548-g31	190.9	0.05	555-g8	138.9	0.08
536-g17	175.9	0.16	548-g32	63.9	0.04	556-g12	106.4	0.17
539-g14	155.7	0.07	548-g50	77.5	0.07	556-g23	148.7	0.05
539-g5	156.0	0.06	548-g63	104.5	0.09	556-g5	166.0	0.07
53-g2	159.9	0.08	548-g71	80.6	0.09	55-g29	169.4	0.12
540-g10	120.7	0.13	548-g77	87.1	0.08	55-g4	245.6	0.03
540-g16	80.0	0.04	549-g18	168.6	0.06	55-g5	78.4	0.19
541-g1	247.0	0.07	549-g22	134.2	0.08	562-g14	199.9	0.04
541-g4	147.8	0.04	549-g40	273.2	0.06	563-g11	269.5	0.08
543-g12	162.2	0.04	54-g21	109.3	0.07	563-g13	182.6	0.05
544-g27	159.2	0.07	550-g7	67.8	0.02	563-g14	148.3	0.02
544-g32	145.1	0.07	550-g9	186.3	0.06	563-g17	244.5	0.07
545-g11	191.3	0.02	551-g13	161.9	0.12	563-g21	356.0	0.03
545-g21	184.0	0.05	551-g16	49.1	0.07	563-g28	193.8	0.09
545-g3	40.6	0.07	551-g31	72.1	0.07	564-g20	81.6	0.04
545-g5	96.6	0.03	552-g43	181.7	0.13	564-g23	175.9	0.05
546-g15	310.4	0.09	553-g26	205.0	0.06	564-g31	171.0	0.06
546-g29	157.3	0.03	553-g3	260.8	0.08	564-g35	109.8	0.04
546-g31	207.4	0.05	554-g10	268.0	0.03	566-g14	179.2	0.07
546-g36	188.1	0.09	554-g19	145.9	0.07	566-g22	136.5	0.03
546-g37	119.1	0.09	554-g24	129.5	0.05	566-g26	188.9	0.04
566-g30	156.4	0.06	576-g51	167.0	0.04	58-g3	148.8	0.13
566-g9	156.0	0.10	577-g1	176.4	0.11	593-g3	186.1	0.05
567-g26	202.7	0.04	579-g25	165.5	0.17	594-g8	186.4	0.07
567-g45	292.9	0.12	579-g9	114.7	0.07	595-g10	132.0	0.05
567-g6	98.2	0.07	57-g80	159.1	0.05	596-g9	126.0	0.11
568-g19	189.7	0.22	580-g29	159.3	0.05	59-g23	169.5	0.05
569-g22	220.4	0.05	580-g37	199.7	0.12	59-g24	252.5	0.04

(Continued...)

Table of galaxies from Persic & Salucci. (Continued)

103

Table of galaxies from Persic & Salucci. (Continued)

name	v_{max} (ks^{-1})	σ (normalized)	name	v_{max} (ks^{-1})	σ (normalized)	name	v_{max} (ks^{-1})	σ (normalized)
570-g2	154.2	0.06	580-g41	104.1	0.05	601-g19	193.4	0.09
571-g12	175.8	0.18	580-g45	130.5	0.06	601-g25	74.9	0.07
571-g15	235.4	0.07	580-g49	134.9	0.06	601-g4	160.1	0.06
571-g16	152.1	0.08	580-g6	160.8	0.05	601-g5	157.2	0.09
572-g18	139.4	0.06	581-g10	126.6	0.12	601-g7	105.0	0.13
572-g22	81.9	0.08	581-g11	188.0	0.11	601-g9	291.0	0.02
572-g49	87.0	0.05	581-g15	173.5	0.11	602-g15	79.7	0.07
573-g14	144.2	0.08	581-g4	120.4	0.07	602-g25	202.6	0.10
573-g6	135.7	0.08	581-g6	118.6	0.09	603-g12	101.0	0.04
574-g28	129.3	0.07	582-g12	166.8	0.06	603-g20	110.3	0.06
574-g32	163.4	0.09	582-g13	245.6	0.11	603-g22	268.8	0.05
574-g33	184.7	0.08	582-g1	230.3	0.32	604-g1	75.6	0.06
575-g53	135.3	0.06	582-g21	214.6	0.07	605-g7	106.7	0.05
576-g11	146.6	0.04	582-g4	99.6	0.10	606-g11	174.1	0.09
576-g12	158.1	0.10	583-g2	208.3	0.06	60-g15	98.5	0.05
576-g14	194.8	0.04	583-g7	398.2	0.07	60-g24	286.0	0.47
576-g26	76.0	0.05	584-g4	211.2	0.07	60-g25	41.2	0.10
576-g32	172.5	0.08	586-g2	119.8	0.07	61-g8	146.5	0.05
576-g39	156.9	0.07	58-g25	198.8	0.06	62-g3	145.1	0.06
576-g3	92.6	0.05	58-g28	76.3	0.08	69-g11	164.0	0.05
576-g48	231.0	0.07	58-g30	177.8	0.04	6-g3	163.0	0.05
71-g14	219.8	0.04	8-g7	152.9	0.05	m-2-2502	167.3	0.04
71-g4	116.9	0.07	90-g9	173.5	0.06	m-2-2-51	278.8	0.10
71-g5	235.8	0.06	9-g10	179.2	0.04	m-2-7-10	130.1	0.05
72-g5	140.4	0.08	holm370	186.9	0.04	m-2-7-33	188.3	0.04
73-g11	218.6	0.07	i1330	225.2	0.07	m-2-8-12	187.6	0.06
73-g22	202.1	0.05	i1474	147.2	0.06	m-3-1042	149.8	0.03
73-g25	145.4	0.07	i2974	238.0	0.03	m-3-1364	165.4	0.04
73-g42	135.5	0.24	i382	201.2	0.05	m-3-1623	196.1	0.05
74-g19	186.9	0.09	i387	248.1	0.12	m-338025	163.0	0.04
75-g37	130.9	0.03	i407	189.2	0.04	n1090	180.8	0.03
79-g14	154.7	0.02	i416	117.9	0.03	n1114	195.3	0.04
79-g3	252.4	0.02	i5078	119.4	0.03	n1163	160.5	0.06
7-g2	152.6	0.15	i5282	207.8	0.08	n1241	282.6	0.07
80-g1	110.8	0.10	i784	191.5	0.05	n1247	268.8	0.03
82-g8	263.0	0.06	i96099	176.1	0.02	n1337	112.9	0.03
84-g10	198.8	0.02	m-1-1035	191.4	0.03	n1417	235.1	0.10
84-g33	288.8	0.04	m-1-2313	158.1	0.04	n1421	170.0	0.19
84-g34	235.2	0.35	m-1-2321	180.5	0.05	n151	325.5	0.05
85-g27	179.2	0.07	m-1-2522	170.1	0.05	n1620	214.0	0.05
85-g2	197.5	0.04	m-1-2524	77.5	0.04	n1752	224.3	0.04
85-g38	178.3	0.06	m-1-5-47	226.0	0.03	n1832	198.7	0.03
85-g61	94.5	0.06	m-2-1009	258.0	0.04	n2584	187.3	0.06
87-g3	283.8	0.16	m-213019	166.5	0.06	n2721	242.7	0.05
87-g50	97.9	0.07	m-214003	150.6	0.04	n2722	134.9	0.05
88-g16	201.4	0.04	m-215006	129.9	0.06	n2763	145.9	0.03
88-g17	350.1	0.15	m-222023	299.8	0.06	n280	319.8	0.05
88-g8	203.9	0.13	m-222025	159.8	0.15	n2980	234.4	0.04
8-g1	113.3	0.08	m-2-2-40	166.1	0.06	n3029	170.4	0.13
n3138	183.0	0.05	n755	133.3	0.03	u12571	184.4	0.05
n3321	144.0	0.07	n7568	224.0	0.05	u12583	105.4	0.05
n3361	136.2	0.04	n7593	141.5	0.06	u14	198.0	0.05
n3456	168.0	0.05	n7606	273.5	0.30	u1938	188.2	0.05
n3715	193.2	0.06	n7631	205.2	0.06	u2020	90.0	0.06
n4348	182.2	0.03	n7677	181.7	0.11	u2079	125.5	0.05
n4705	195.7	0.02	u12123	118.9	0.06	u210	111.3	0.06
n697	197.2	0.02	u12290	240.0	0.05	u321	79.8	0.07
n699	200.9	0.04	u12370	116.8	0.09	u541	118.2	0.07
n701	125.3	0.11	u12382	124.0	0.10	ua17	109.9	0.02
n7218	128.4	0.02	u12423	262.6	0.07			
n7300	244.7	0.03	u12533	253.8	0.04			
n7339	156.3	0.02	u12555	114.8	0.07			
n7536	183.6	0.04	u12565	186.3	0.09			

This table lists fitted v_{max} and normalized σ of fit.
Index to names are n: NGC, m: Messier, i: IC, u:UGC, holme: Holmberg, others are ESO numbers.

Table 12.4: Comparison of Distances - Roxy's Ruler vs Cepheids.

Name	v_{max} km/s	α (arcmin)	Equ (12.22) distance (Mpc)	Norm. Error	m-M	Cepheid distance (Mpc)	Norm. Error	Ref
NGC 7331	22.5	1.15	12.1	0.20	30.89	15.1	0.15	1
NGC 3319	13	1.94	12.4	0.12	30.78	14.3	0.15	2
NGC 4321	13	1.68	14.26	0.16	31.04	16.2	0.14	3
NGC 4414	23	0.66	20.7	0.33	31.41	19.2	0.15	4,5
NGC 224	24.1	18.94	.7	0.19	24.44	.8	0.15	6
NGC 3627	19	1.55	10.6	0.28	30.06	10.3	0.26	7
NGC 4536	12.5	1.41	17.7	0.13	30.95	15.5	0.11	8
NGC 3031	14	5.77	3.9	0.11	27.8	3.6	0.12	9
NGC 3351	22	1.53	9.3	0.23	30.01	10.1	0.12	10
NGC 2090	15	1.96	10.6	0.14	30.45	12.3	0.12	11
NGC 4548	15.7	1.59	12.5	0.23	31.04	16.2	0.12	12
NGC 925	12	2.18	11.9	0.12	29.84	9.3	0.12	13
NGC 3198	15.1	1.35	15.15	0.15	30.8	14.5	0.09	14
NGC 4639	20	0.73	21.2	0.17	31.8	22.9	0.14	15
NGC 4725	21	1.2	12.4	0.13	30.57	13.0	0.12	16
NGC 3368	22	1.44	9.9	0.51	30.2	11.0	0.15	16
NGC 5457	19	2.29	7.2	0.11	29.34	7.4	0.15	16
NGC 598	13	33.8	.7	0.10	24.64	.8	0.14	16,17
NGC 4535	14	1.24	18.0	0.12	31.1	16.6	0.11	17
NGC 1365	5	3.4	18.4	0.11	31.39	19.0	0.15	18
NGC 2541	9.5	2.39	13.8	0.15	30.47	12.4	0.12	19

References. (1) [Rubin & Ford (1970)]; (2) [Moore & Gottesman (1998)];
(3) [Knapen et al (1993)]; (4) [Braine et al (1993)];
(5) [Vallejo et al (2002)]; (6) [Abell (1975)];
(7) [Chemin et al (2003)]; (8) [Afanasev et al (1991)];
(9) [Rohlfs & Kreitschmann (1980)]; (10) [Devereaux et al (1992)(];
(11) [Kassin & Weiner (2006)]; (12) [Vollmer et al (1999)];
(13) [Pisano et al (1998)]; (14) [Begeman (1989)];
(15) [Rubin et al (1999)]; (16) [Brownstein & Moffat (2006)];
(17) [Woods et al (1990)]; (18) [Lindblad et al (1996)];
(19) [Józsa (2007)]

12.7   Mass, Linear Density and Luminosity

(This part was written by Cam. I'm so proud.)

In the model we are presenting, gravitationally self bound particles are oriented along the path of the spiral shaped geodesic as in equation 12.20. An examination of the presented model yields a straightforward transformation into L-1 space and using the Lebesgue, [Lebesgue (1902)], measure of linear density in a singular dimension, we have:

$$\rho_l = \frac{v_{max}^2}{2G}$$
(12.26)

This matches the measurements of Fish, [Fish (1961)], in which it was found that galaxies have a luminosity profile of a structure having a constant radial luminosity. Fish used a series of concentric rings measuring the luminosity determined by photometry centred on NGC 5055 to discover this relation. We repeat this measurement on photographs of M 101, with comments, in a following section.

Using previously described distance measures and the angular length of the major axis, the intrinsic major axis length, L, can be determined. Using equation 12.35 we can determine the mass of the galaxy as:

$$M_g = L\rho_l \tag{12.27}$$

From this we can calculate the angular momentum of a galaxy as:

$$l = v_{max}\rho_l L^2 \tag{12.28}$$

12.7.1 Using Luminosity to Measure Density

A photograph of a galaxy is a two dimensional projection of light from a three dimensional cloud of stars. For such a projection, the variances in luminosity could reflect the variances in mass if certain assumptions are made regarding the ratio of luminosity to mass if there are no irregular distributions of undetectable material and stars of different light/mass ratio are distributed evenly along the radial dimension. M 101 is an interesting galaxy where young and old stars are distributed rather evenly throughout the galaxy, [Suzuki (2007)]. More specifically: "The distribution of the cold dust is mostly concentrated near the center, and exhibits smoothly distributed over the entire extent of the galaxy, whereas the distribution of the warm dust indicates some correlation with the spiral arms, and has spotty structures..."

The L/M ratio is expected to be higher in areas of warm dust due to the formation of young bright stars. Since the radial arms are distributed throughout the galaxy in the radial direction, for the analysis of M 101 it can be assumed that the L/M ratio is constant for any spherical surface of radius r. Under this assumption, the presence of any irregular distributions of undetectable material will be apparent as a discrepancy between the luminosity curve and the linear mass function.

For a cloud with a constant rotation profile, the mass subtended at radius r must be linearly proportional to r. If young and old stars are distributed such that the M/L ratio for M 101 is the same for any spherical surface of radius r from the center, then the presence of any non-luminous material will become evident if the luminosity curve $L(r)$ does not correlate to that expected of a linear mass function.

Considering a galaxy as a cloud comprised of stars and particles in three dimensions, the density of the galaxy subtended by a sphere of radius r is given as:

$$\rho(r) = \frac{M(r)}{V(r)} \tag{12.29}$$

Where,

$\rho(r)$ = the density of the cloud contained by a sphere of radius r

$M(r)$ = the mass of the cloud contained by a sphere of radius r

$V(r)$ = the volume of a sphere of radius r

If material is in circular orbit about a mutual centre we have:

$$a = \frac{v^2}{r} \tag{12.30}$$

$$ma = m\frac{v^2}{r} \tag{12.31}$$

$$F_g = m\frac{v^2}{r} \tag{12.32}$$

$$G\frac{M(r)m}{r^2} = m\frac{v^2}{r} \tag{12.33}$$

$$G\frac{M(r)}{r} = v^2 \tag{12.34}$$

$$\rho_l = \frac{v^2}{2G} \tag{12.35}$$

M 101 has a tangential rotation profile of approximately 250 K s^{-1}, [Comte (1978)]. From Equation 12.35, the distribution of matter orbiting a mutual center of mass with the same tangential velocity, is given as:

$$M(r) = kr \tag{12.36}$$

Where,

$M(r)$ = the mass of all stars subtended by a sphere of radius r

k = some constant

r = radial distance from the center of the galaxy

For spiral galaxies with constant rotation profiles:

$$\rho(r) = \frac{kr}{\frac{4}{3}\pi r^3}$$

$$= \frac{3k}{4\pi}\frac{1}{r^2}$$

(12.37)

and,

$$\rho_s \propto \frac{1}{r^2}$$

(12.38)

Where, ρ_s = The density of a spiral galaxy with constant rotation profile expressed in units of $\frac{M_\odot}{ly^3}$.

Two Dimensional Projection

Projecting onto two dimensions, such as in Figure 12.26, the area density subtended by r on the surface of the photograph is:

$$\rho' = \frac{M(r)}{A(r)}$$

(12.39)

$$\rho' = \frac{kr}{\pi r^2}$$

(12.40)

$$\rho' = \frac{k}{\pi r}$$

(12.41)

Where, $\rho' G_s$ = the area density of a 2D projection of a spiral galaxy with constant rotation profile expressed in units of $\frac{M_\odot}{ly^3}$.

The mass of any material between two concentric circles on a photograph

$$M(\Delta r) = \frac{k}{\pi}\left(\frac{1}{r_2}A_2 - \frac{1}{r_1}A_1\right)$$

(12.42)

$$M(\Delta r) = \frac{k}{\pi}\left(\frac{1}{r_2}\pi r_2^2 - \frac{1}{r_1}\pi r_1^2\right)$$

(12.43)

$$M(\Delta r) = k\left(r_2 - r_1\right)$$

(12.44)

$$M(\Delta r) = k\Delta r$$

(12.45)

108

(a) Blue Image POSSII-J Sky Survey Filter: GG395 > 400nm

(b) Red Image POSSII-F Sky Survey Filter: RG610 > 600nm

(c) IR Image POSSII-N Survey Filter: RG9 > 750nm

(d) Blue Image Linear Reduction Grey Value -23/255

(e) Red Image Linear Reduction Grey Value -23/255

(f) Near-IR Image Linear Reduction Grey Value -23/255

Figure 12.26: Reducing the grey value uniformly to correct for noise, film fog, background light, and foreground light. The STScI Digitized Sky Survey provided digitized scans of the above three images of M 101 with vertical and horizontal separation of 30 arcseconds.

For any concentric spheres of r_1 and r_2 where $r_1 - r_2 = \Delta r$, the same mass will exist between the two spheres when Δr is the same.

We present Figure 12.27 in which two different geometrical orientations of matter show distinctive theoretical profiles. Luminous matter (stars) oriented in the shape of a disc would show a luminosity profile of a line having some non-zero slope through the origin. A linear orientation of stars would show a luminosity profile of a horizontal line. In Figure 12.28 we see the two straight lines and the measured relative luminosity from digital photographs of M 101 indicating a match with a linear orientation of stars rather than that of a disk.

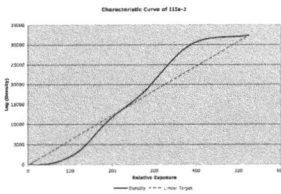

(a) light density against relative exposure

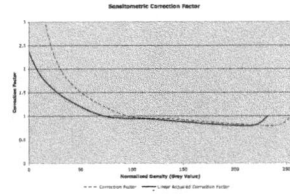

(b) Correction factor over normalized density

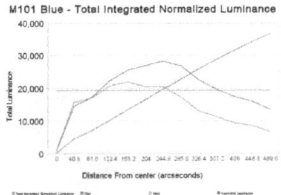

(c) Blue filter luminance of M 101

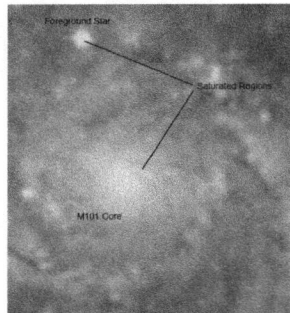

(d) Foreground star and galactic core

Figure 12.27: Total luminance decreases it's value drastically closer to the center due to saturation of the emulsion. The original image files show a region of aberration on the original images in the brightest regions. This pattern is assumed to be due to saturation of the IIIa-J film. Note the presence of a foreground star near the centre of M 101

110

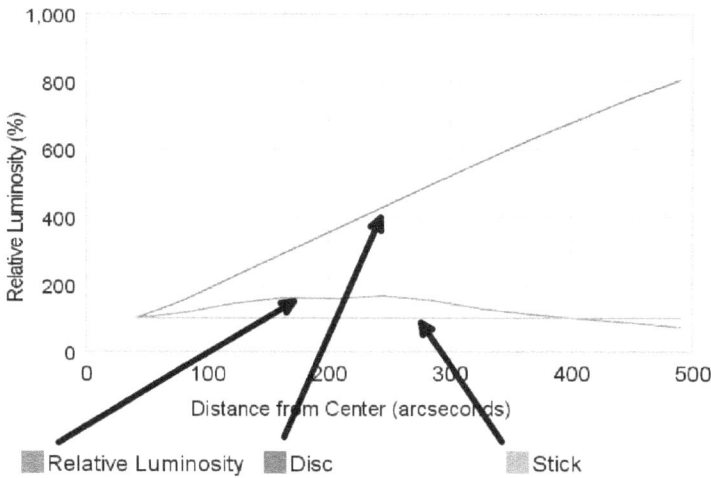

Comparison of Relative Luminosity Curves for Possible Galactic Geometries

Figure 12.28: The relative luminosity profile of M 101 and the comparative theoretical luminosity profiles of a disk and a linear orientation. The measured relative luminosity shows a very close match to a linear orientation of stars.

12.7.2 Using the Length of the Bar, in Barred Spirals, to Validate the Model

Equation (12.12) describes a non-linear function between $\gamma(r)$, the Lorentz factor and r, the distance from the centre of the galaxy. It is smooth, with $\gamma(r)$ having a value of very nearly 1 until r reaches a value of c/ω_0. A geodesic is usually a straight line in most circumstances, unless there is some fairly high degree of curvature in the space-time continuum for it to not be "straight" in the usual way that such a concept is thought of. Although a straight line can be thought of as the path of light, or a line of sight, perhaps it can also be thought of as a line having a constant direction. If this is the case, then a radial geodesic would appear as "straight" in a large rotating coordinate system from the centre to a distance of c/ω_0. Further from the centre, this geodesic would appear as a spiral as described. Some galaxies appear as barred spirals. Since we know the distance to galaxies, we can measure the length of the apparent bar and compare that to the length of c/ω_0 for each galaxy. We have estimated the lengths of bars in six barred spirals by measuring the lengths of the bar in pixels and adjusting for the estimated angle of incline of the galaxy and the angle of orientation of the bar to the major axis. These measurements have a high degree of error since the length of a bar is not an exact measurement for all barred spirals. Some barred spirals, such as NGC 1365, have a well defined bar with a distinct transition region between bar and spiral areas. Other barred spirals, such as NGC 925, have a large transition region and there is an indistinct transition between the bar and spiral. Nevertheless, some estimate can be made and reasonable error estimates can be presented. We list these galaxies in Table 12.5. The first column is the name of the galaxy, the second is the measured length of the bar in Kpc, the third is the sigma of the measure, the fourth is the predicted length of the bar in Kpc and the fifth is the sigma of the prediction based on v_{max} and distance from equation (12.22) according to Table 12.4. Two figures are also presented. Figure 12.29 shows the predicted bar length vs the measured bar length and Figure 12.30 shows the predicted curve of bar length vs v_{max} as well as the measured bar lengths. The fitted slope through the origin of Figure 12.29 is 0.943 and the sigma standard deviation of the comparison is 1.51 Kpc. Even though there is a high allowance for error, there is still a valid correspondence between bar length and $1/\bar{\omega}_0$ as derived in the model.

Table 12.5: Bar Length of Barred Spirals

Name	Length of bar (Kpc)	σ (Kpc)	Predicted length of bar (Kpc)	σ (Kpc)
NGC 1365	10.9	6.02	11.8	1.18
NGC 925	7.4	4.17	4.9	0.49
NGC 4536	6.1	3.48	4.7	0.47
NGC 3319	3.9	2.19	4.5	0.45
NGC 4535	2.7	1.53	4.2	0.42
NGC 4548	4.6	3.11	3.8	0.38

12.8   Conclusions

The Fundamental Theorem of Algebra demands a unique and existent solution to the problem of modelling a body consisting of many particles, each having a circular orbit about a common centre, each have approximately the same mass, and the body held together by mutual gravitational attraction between the particles making up the body entire. The derivation is very straightforward and rather

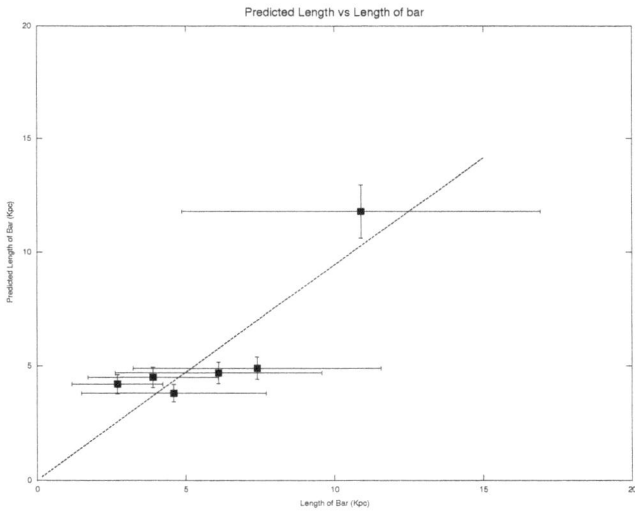

Figure 12.29: Graph comparing the predicted bar length vs the measured bar length. The slope of the fitted line is 0.94293 and the standard deviation of the comparison is 1.52 Kpc.

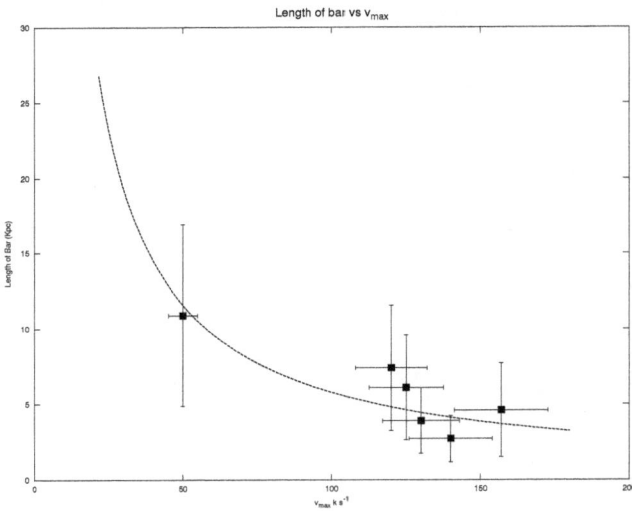

Figure 12.30: This is a graph of the bar spirals presented. The graph shows bar length vs v_{max}. The curved line shows the theoretical length as a function of v_{max} which appears to be a nice fit to the data.

simple. However, the derivation of linear density began with a Newtonian non-relativistic approach. This approach does not take into account the finite value of the speed of light or time delay in gravitational influences. Once the gravitational influences have perpetrated to some other body, the body from which the gravitational attraction has emitted, has moved. Over long distances, this can be significant. It is much easier in calculating the mathematical requirements of this model to begin with a Newtonian approach to gravitational influences so long as an understanding is firmly held in mind that the calculations must be rectified to take into account Minkowski, Lorentz and Einstein's theorem of time and space dilation. (Or at least the measurements thereof). A physical model which can be brought to mind is that of a stick of constant density made up of stars and held together by mutual gravitational attraction between the stars making up the stick. This analogy is comparable to an ordinary stick made up of molecules whereby the stick is held together by the mutual electrostatic attraction between the molecules making up the stick. In both cases the two sticks are held together by mutual inverse-square forces. The difference between them is merely a matter of scale. The electrostatic attractive forces are much stronger than gravitational forces by about 40 orders of magnitude. However, even though stars are distributed light years apart, the overall linear density of a galaxy is about 10^{20} Kg per meter. That is a very dense stick indeed and has sufficient mass per unit length, or linear density, to keep it "stuck" together. If an approach is taken that there is a gravitational force contracting the stick and the gravitational force being inverse square, or absolutely convergent, then this approach would meet requirements for being in a non-Euclidean, or L-1, space. Without this model, that of a stick of stars rotating in space, a pseudo-mass to create pseudo-gravity would be required to explain the behaviour of NGC 3198. Instead, the application of general relativity to this system matches the rotation parameters of NGC 3198, explains the spiral shape of the galaxy and correctly predicts its flat velocity rotation profile. It is a stick in L-1 space and a spiral in either L-2 or L-4 space.

If the stars of NGC 3198 were in any orbit other than circular and in a single plane, the galaxy would not maintain its distinct spiral shape. Because there are many other spiral shaped galaxies with a morphology similar to that of NGC 3198, it is not unreasonable to believe this spiral morphology is somewhat stable. Therefore, we must conclude that galaxies are rotating as rigid bodies. It is the only thing that makes sense. The orbital speed of material in a galaxy is determined by measuring the shifting of spectral lines of sections of the galaxy. It is not determined by the distance travelled by stars divided by the time it takes for stars to get farther along in its orbit. The distances the stars are travelling are so vast, and we have not observed galaxies for a long enough time, that we have not been able to detect circular motion directly. The shift in spectral lines

115

may not, and indeed cannot, be the result of the distance travelled divided by time measured using our own local clocks and rulers here on Earth. Things must be thought of locally, meaning in the distant local region of a star orbiting its home galaxy. And in the galaxy's stars' local reference frames, the stars are not moving at all; each star is stationary relative to itself. However, the clocks and rulers in each local reference frame differ from the clocks and rulers of different stars in the galaxy due to the variation of the Lorentz transformations at different radial distances from the centre. The clocks of each Hydrogen atom in this distant galaxy vary from our local clocks since the distant orbiting atoms are in a region of curved space and time. Therefore, the shifting of the spectral line emitted or absorbed by a Hydrogen atom is affected by both the Lorentz transformation of Doppler effects and from the Lorentz factor resulting from the curvature of space-time in a large rotating system. The Lorentz-Doppler effect varies directly with r and the Lorentz-curvature effect varies inversely with r. The two effects cancel in outer regions of the galaxy and this results in a constant velocity rotation profile.

We therefore conclude:

1. NGC 3198 is gravitationally self bound having a linear mass distribution function of constant linear density.

2. There is no requirement for exotic material in the galaxy's mass distribution to explain the galaxy's rotation curve or its luminosity profile. There is no gravitationally detectable "dark matter" in NGC 3198.

3. Einstein's general theory of relativity and Newton's principles of gravitational attraction hold over very great distances.

Figure 12.31: The central region of our galaxy. There is no spiral arm structure visible since we are looking down a geodesic into the centre. The geodesic is a line of sight straight down into the photograph like looking into a deep well. The central region of the galaxy here appears at the bottom of the well. By the time the light and gravitational influence of the stars in the central region have reached us they have all moved in their respective orbits. In our reference frame, there appears to be a ball of stars surrounding the central region. This ball of stars is pulling us towards the centre of the galaxy while our tangential rotational velocity keeps us in a circular orbit, [Henley]

.

Chapter 13

An Unlucky Chapter

This chapter has been left deliberately blank except for this notice which means that the chapter is not completely blank but almost so.

Chapter 14

111 Spiral Galaxies

A previous chapter showed a way to determine the distance to galaxy NGC 3198, This was determined from its spiral morphology and rotational velocity. It was also seen that the stars of the galaxy appear to orbit the galaxy with the same common orbital velocity independently of the distance from the centre of the galaxy. This is only possible if the member stars of spiral galaxies behave gravitationally and visually as though galactic matter, notably stars and interstellar dust and gas, have a linear orientation with constant linear density. However, spiral galaxies have a spiral morphology which is the result of the finite velocity of light and gravitational interactions over very large distances coupled with a relatively high rotational velocity. This is an excellent example of delayed gravitational interaction and the Lorentz transformations applied to a rotating reference frame. Each star orbiting within the galaxy detects all of the other stars in a stationary linear orientation. Observers not orbiting the galaxy, and in a comparatively inertial reference frame, see the orientation of stars as a spiral. Since the relativistic time/distance dilations are known; the absolute size of the galaxy can be calculated and a direct measure of its distance can be made.

The spiral shape of the galaxy is determined by:

$$r = \frac{2\pi\theta}{(v/c)} \tag{14.1}$$

where r is in light years and θ is in radians. From this we can determine the distance to the galaxy by:

$$\Delta r = \frac{2\pi^2}{(v/c)}$$

$$d = \frac{\Delta r \times 360 \times 60}{4\pi\alpha} \qquad (14.2)$$

$$d = \frac{2\pi^2 \times 360 \times 60}{(v/c)4\pi\alpha}$$

where Δr is the distance between spiral arms along the major axis of the galaxy in light years, α is the angular separation between spiral arms along the major axis of the galaxy in minutes of arc, v is the rotational velocity of stars in the galaxy, c is the speed of light and d is the distance to the galaxy in light years. The above distance formula to galaxies is known as "Roxy's Ruler".

The rotational velocities of galaxies studied were originally taken from a survey of 1,355 galaxies in the southern hemisphere by Mathewson, [Mathewson et al (1992)]. We have included a typical page of Mathewson's rotation profiles, Figure 14.1, to show that the member stars of spiral galaxies orbit the galactic centre with a common rotational velocity. It can be seen that each of the velocity profiles show lines of constant rotational velocity extending to the ends of the graphs, one showing the receding arm and the other showing the advancing arm. There is an adjoining line between these two extensions crossing the central location of each galaxy. The shape is somewhat like a pulled apart "Z". We interpret this as a horizontal line representing one side of the galaxy where all the member stars have the same velocity, say, of recession, and the opposite side of the galaxy where stars are approaching us, as in this example. The spectral line is measured with a slit, or beam of a radio telescope, across the galactic plane. This causes the line portraying the measurements in the central region to be diagonal.

We have used the analysis and results of Persic, [Persic & Salucci (1995)], for rotation velocities of each galaxy which is presented here. The data has been corrected for angle of incline for each galaxy and transformed into a heliocentric reference frame.

We have measured the distances to 111 galaxies in the southern hemisphere using the above described method. Since the distance to each galaxy is known, we can then measure its total length, mass and angular momentum. This data is presented in the following table of galactic data. Various graphs are displayed as well.

We note from figure 14.2 that there appears to be no relationship between galactic red shift and distance. The analysis shows a statistical $\mathbf{R}^2$, or coefficient of determination, of -1.0558, (less than zero, actually less than negative one), upon an attempted linear fit. There is, therefore, no acceptable linear fit and the data appears completely random. However, it is obvious that all of the galaxies displayed are indeed red shifted. It appears unlikely, therefore, that this red shift is the result of a Doppler effect.

We call the rotational velocity of the stars within the galaxy the spin velocity or spin of the entire galaxy.

We can also see that:

1. more massive galaxies spin faster

2. more slowly spinning galaxies tend to be more spread out and therefore larger than faster spinning ones

3. spiral galaxies have similar masses to within an order of magnitude

Disscussion Concerning Error

The distance to the galaxies are determined from measurements of angular distance between galactic arms along the major axis of the galaxy and their spin velocity. Mathewson reports an error within 10 Kps in spin velocity and upon examining the data, we feel that this is acceptable. The angular distance between galactic arms is determined from noting the pixel locations at either end of each galaxy on well defined digital photographs from the Hubble space telescope. We submit an error estimate of 2 pixels for each measurement. The angular width along the major axis of each galaxy was taken from the Simbad data base. This angular width, combined with the described distance measure, yields the mass of each galaxy and resultant angular momentum. The linear density of a galaxy is given as, $\rho_l = v^2/(2G)$.

Table 14.1: Table of Galactic Data.

Name	Rotation Velocity (Kps)	Dist (MPc)	Red Shift (Kps)	Length (10^3) LY	Mass (10^{11} Solar Masses)	Angular Momentum (10^{67} J-s)
1-G6	137	49	2245	139	0.93	3.34
1-G7	120	158	4994	176	0.9	3.6
101-G20	178	97	5845	172	1.94	11.16
101-G5	178	103	6638	154	1.74	9.01
102-G10	178	74	4698	161	1.82	9.84
102-G15	178	104	5018	188	2.12	13.35
102-G7	227	99	5014	82	1.5	5.22
103-G13	210	32	4664	71	1.11	3.12
105-G20	122	84	5672	138	0.73	2.32
105-G3	162	72	4860	103	0.97	3.05
106-G12	130	103	4155	145	0.87	3.09
107-G36	208	23	3096	74	1.14	3.29
108-G11	214	97	2979	183	2.99	21.98
108-G19	165	47	2956	76	0.73	1.72
113-G21	90	107	4822	136	0.39	0.91
114-G21	166	101	6378	123	1.21	4.67
116-G14	152	55	5417	91	0.75	1.94
117-G18	206	80	5795	98	1.48	5.6
117-G19	177	58	5386	106	1.19	4.2
120-G16	138	71	3674	120	0.81	2.54
121-G26	226	34	2220	94	1.72	6.87
121-G6	146	33	1228	125	0.95	3.26
123-G15	232	45	3215	121	2.32	12.24
123-G16	100	84	3194	144	0.51	1.4
123-G23	160	46	2910	100	0.91	2.76
123-G9	151	63	3183	117	0.95	3.17
140-G24	206	76	3183	117	1.77	8.02
140-G25	100	76	2047	194	0.69	2.52
109-G32	112	85	3362	137	0.61	1.78
116-G12	145	42	1153	141	1.06	4.06
140-G28	111	100	4875	150	0.66	2.06
140-G34	103	70	3405	100	0.38	0.73
141-G20	238	51	4349	100	2.03	9.1
141-G34	271	50	4404	139	3.64	25.8
141-G37	282	50	4386	81	2.31	9.98
141-G9	219	40	3636	125	2.13	10.94
142-G30	181	70	4201	115	1.34	5.26
142-G35	223	44	2031	139	2.46	14.36
145-G22	198	71	4465	104	1.45	5.62
146-G6	108	110	4598	165	0.69	2.31
151-G30	180	91	5335	140	1.61	7.64
155-G6	105	24	1070	90	0.35	0.63
162-G15	93	205	2839	281	0.87	4.26
162-G17	60	128	2839	253	0.32	0.93
163-G11	198	45	2839	102	1.42	5.41
18-G13	219	58	2839	114	1.95	9.18
183-G5	90	117	2839	167	0.48	1.37
184-G51	230	69	2839	142	2.68	16.46
184-G54	160	44	2839	62	0.57	1.05
184-G63	170	57	2839	124	1.28	5.09
184-G67	242	43	2839	126	2.62	15
185-G36	155	71	2839	104	0.89	2.7
185-G68	114	94	2839	132	0.61	1.72
185-G70	133	79	2839	99	0.62	1.54
186-G21	188	90	2839	144	1.82	9.3
186-G75	180	86	2839	126	1.46	6.24
186-G8	141	93	5709	136	0.97	3.49
187-G6	105	95	4652	150	0.59	1.74
187-G8	121	130	4404	182	0.95	3.93
196-G11	116	51	3637	79	0.38	0.65
197-G2	172	131	6306	192	2.02	12.54
197-G24	157	113	5877	178	1.57	8.24
200-G3	105	68	1034	249	0.98	4.82
202-G26	134	69	5111	93	0.59	1.39
204-G19	122	79	4516	121	0.64	1.79
208-G31	155	59	3068	99	0.85	2.45
215-G39	140	104	4335	136	0.95	3.42
216-G21	181	49	5086	65	0.76	1.68
220-G8	145	61	3013	110	0.82	2.46
231-G23	230	67	5024	111	2.08	9.97
233-G36	116	118	3291	155	0.74	2.51
233-G41	267	45	2951	106	2.69	14.3
233-G42	86	189	2561	253	0.67	2.72
234-G13	135	77	3186	108	0.7	1.93
235-G16	196	48	7147	64	0.88	2.09
235-G20	150	105	4671	172	1.38	6.7

(Continued...)

124

Name	Rotation Velocity (Kps)	Dist (MPc)	Red Shift (Kps)	Length (10^3) LY	Mass $(10^{11}$ Solar Masses	Angular Momentum $(10^{67}$ J-s)
236-G37	180	57	5558	84	0.97	2.74
237-G49	87	158	2913	343	0.92	5.18
238-G24	209	72	7013	101	1.57	6.19
240-G11	235	31	2876	143	2.82	17.9
240-G13	143	70	3267	87	0.63	1.49
243-G8	174	85	7323	120	1.29	5.06
244-G31	242	48	6726	70	1.47	4.69
244-G43	160	79	6231	96	0.88	2.56
249-G16	186	27	1179	194	2.39	16.19
25-G16	125	113	6136	115	0.64	1.72
250-G17	261	25	4541	60	1.45	4.27
251-G10	230	53	4451	89	1.68	6.51
251-G6	142	107	4981	147	1.06	4.16
265-G16	174	84	5166	144	1.55	7.33
266-G8	113	118	3225	178	0.81	3.05
267-G29	200	66	5445	90	1.28	4.33
267-G38	225	81	5884	97	1.75	7.19
268-G11	231	31	8517	32	0.6	0.82
268-G33	215	49	5502	97	1.6	6.32
269-G63	146	100	3189	154	1.17	4.94
27-G24	200	106	4079	133	1.89	9.44
124-G15	129	78	2606	129	0.76	2.39
2-G12	137	101	4643	138	0.92	3.28
22-G3	107	131	2737	216	0.88	3.83
231-G29	113	88	4940	136	0.62	1.78
249-G35	50	197	1035	282	0.25	0.67
26-G6	110	75	2743	166	0.72	2.47
269-G15	149	76	3376	193	1.53	8.24
269-G19	189	37	2173	154	1.96	10.72
269-G49	94	112	3238	153	0.48	1.3
269-G61	247	44	4917	105	2.27	11.05
284-G21	150	69	5773	75	0.6	1.27
285-G40	240	73	6735	110	2.25	11.13
286-G18	303	46	9150	115	3.77	24.75
287-G13	172	40	2703	86	0.9	2.51

Table of galactic data. (Continued)

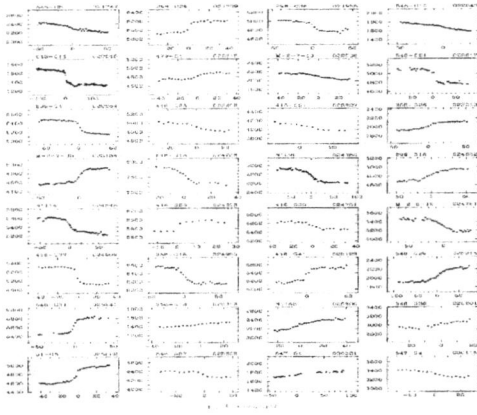

Figure 14.1: A typical page from the Mathewson Survey.

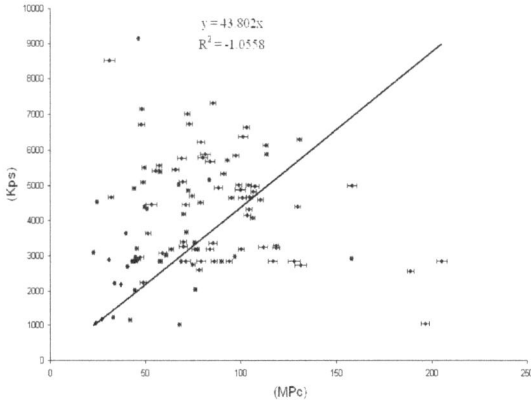

Figure 14.2: This is what is known as a Hubble diagram. On the left, or "y", axis is the red shift of galaxies which is supposed to denote the speed of recession of the galaxy. Across the bottom, or "x", axis, is the distance to the galaxy. Using Cepheid variables, we only get to see about 22 MegaParsecs distance. With Tulley-Fisher, we only get to see up to about 32 MegaParsecs distance. This graph shows a distance to about 220 MegaParsecs. If the universe is expanding, this graph should be a straight line. The confidence variable, R^2 is so bad it is negative. This is a Hubble diagram of 111 galaxies in the Southern Hemisphere with error bars for distance measurements. Red shift is in Kps and distance in MPc. There appears to be no linear relationship of red shift to distance. If anything, this proves conclusively that the universe is not expanding.

127

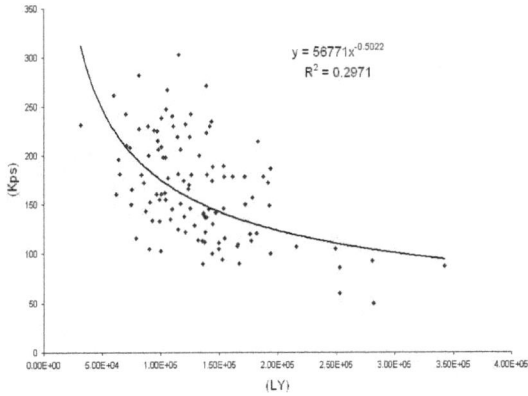

Figure 14.3: Rotational velocity of stars of galaxies vs. overall length of the galaxy in the reference frame of the member stars.

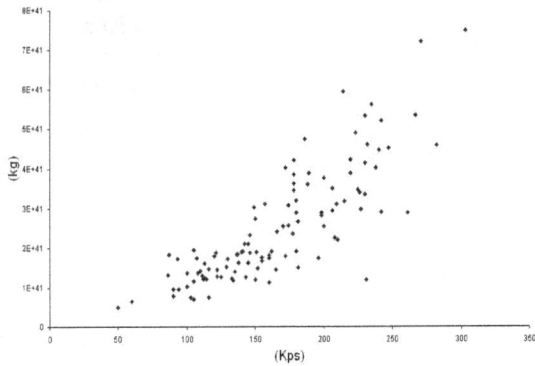

Figure 14.4: Mass of each galaxy vs. rotational velocity of stars within the galaxy.

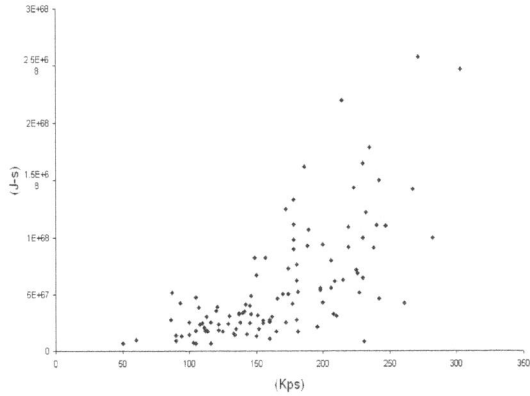

Figure 14.5: Angular momentum of each galaxy vs. rotational velocity of stars within the galaxy.

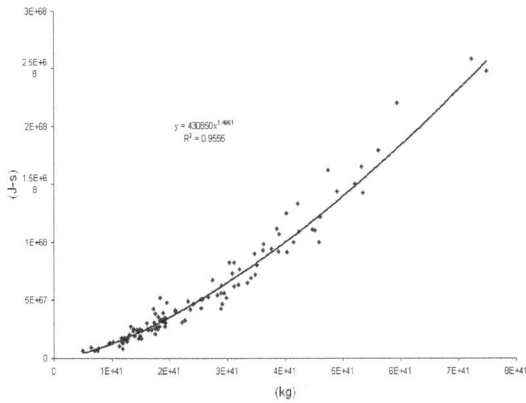

Figure 14.6: Angular momentum of each galaxy vs. mass of galaxy.

Chapter 15

Entropy Reversal
at Galactic Centres

Consider the thermodynamic equation of a spinning black hole with charge from Brandon Carter, Stephen Hawking and James Bardeen, [Carter et al (1973)]:

$$dM = \frac{\kappa}{8\pi} dA + \omega dJ + V dQ$$

where M is the mass, κ relates to the temperature, A is the surface area, ω is the angular velocity, J is the angular momentum, V is the electrostatic potential and Q is the charge. Each of the terms are forms of energy: The mass energy, (total energy) equals the heat energy plus the energy of angular momentum plus the electrical energy. However, to make all the units agree, let us multiply c^2 to the left to make:

$$c^2 dM = \frac{\kappa}{8\pi} dA + \omega dJ + V dQ$$

The heat energy term, $\frac{\kappa}{8\pi} dA$ can be thought of as the temperature of the black hole times the surface heat capacity times the area. Or:

$$\frac{\kappa}{8\pi} dA = C_A T dA$$

And looking at the angular momentum we have:

$$\omega dJ = \frac{v}{R} R \times v dM$$

131

$$\omega dJ = v^2 dM$$

where $J = R \times vM$ and v is the tangential velocity of the black hole at its equator. Let us say the black hole is spinning but electrostatically neutral. Rewriting:

$$\frac{1}{C_A} \frac{dM}{dA} (c^2 - v^2) = T$$

and here we see that as v approaches c the temperature of the black hole decreases and when $v = c$ the temperature of the black hole is absolute zero.

At first it may seem that certain inviolable laws of Physics have been broken, (which is true), but that is not really important. Let's examine this step by step. Let us say there is a black hole and stuff falls into it increasing its spin. Sooner or later it is spinning so fast that the equator is moving at the speed of light. Please remember that the surface of a black hole is not a material object, it is a region where the measure of the escape velocity is the speed of light. We cannot measure, detect or discuss anything within this radius; however, we can discuss what happens at the surface of a black hole. The surface is "made up" of a measure and we shall denote the clocks and rulers involved in that measure as space-time. It is convenient to consider space-time as having certain properties similar to a fluid which has, in itself, a natural speed of interaction, i.e. the speed of light, and a Young's modulus, as a result of the stresses and strains within a gravitational field. It could also have stresses and strains from an electric and magnetic field, however we will restrict our discussion to gravitational fields. These measures are moving, or spinning, which denotes the spin of the black hole. The angular momentum is calculated from the spin of the coordinate system at the surface of the black hole. Once the surface of a black hole at its equator has reached the speed of light, it cannot spin any faster.

An important property of space-time is that it cannot move faster than the speed of light. It cannot even appear, or be measured, to move faster than the speed of light. Of course, what is moving is the coordinate system, which is a special coordinate system: the coordinate system cannot move faster than the speed of light and every point in the coordinate system measures exactly the same speed of light as any other point in the coordinate system[1]. Furthermore, there are circumstances in which the space-time continuum is under tension, strain and shear in order to compress or dilate clocks and rulers in order to maintain this important dictum. We ask that the reader consider that space-time is a medium which can transport electromagnetic radiation as disturbances moving at the speed of light within it. It can also carry gravitational influences in a similar way; however gravitational disturbances involve tidal effects as well as

---

[1]That is pretty well the definition of a Minkowski space

132

stresses and strains along the field lines. Gravitational influences also involve shears in the off-diagonals of both the Einstein and Stress-Energy tensors. These stresses, strains and shears are often denoted as forces between bodies and are described in the Stress-Energy tensor. The alteration of the corresponding Einstein tensor from having zero-valued components is what I have come to call curvature, in that it deviates from what is understood as flat space-time. This curvature can also be linked to terms of acceleration. Linking curvature to acceleration is known as the Principle of Equivalence. That is that an accelerating reference frame is equivalent to a gravitational field.

Once a black hole is spinning so fast that its equator is moving at the speed of light it has no further degrees of freedom. It is at absolute zero. The surface of the black hole is not a material object. It can be thought of as a set of coordinate lines which are capable of moving at speeds up to and including the speed of light. They can also be so wound as to reach the thermodynamic limit of absolute zero. We present a derivation of the Einstein tensor describing sets of coordinate lines on the surface of a black hole. We then present the consequent measures of these lines at the surface of a black hole whose equator is moving at the speed of light and follow through by throwing a brick into it and then mathematically predicting certain results which can be verified through observation.

Techniques for mathematically deriving the Einstein Field Equations can be found in the literature. What we present here is a derivation and explanation of the Einstein tensor on the spherical surface of a black hole. It may be that those in the field of observational astronomy may be unfamiliar with or have not seen the usefulness and ease with which tensors can work in simple geometries. Let us begin with basic principles and develop step by step the geometry of a curved manifold in a four-dimensional Minkowski space.

15.1   The Virtual Observer

Both Werner and I were working on things like the surface of some manifold and we derived a humorous concept which has come in very handy: the virtual observer. The virtual observer is a woman physicist/mathematician, holding a clipboard and pencil, wearing a lab coat and sitting on a chair. Both she and the chair are arbitrarily small or large. She can detect all of her surroundings and is incredibly, incredibly intelligent. She has access to all the knowledge of the universe and can detect anything. And, she does not affect her surroundings in any way. She can be anywhere and move anywhere instantaneously and go back and forth in time however she wishes. She knows everything.[2]

If we are far from the surface of a black hole and watch a clock tied to a brick being thrown into it, we notice that as the brick approaches the surface of the black hole both the brick and clock slow down and eventually freeze at the surface of the black hole, taking an eternity of time to get there. This is from the Schwartzchild metric which determines measures due to the curvature of space and time for a not-spinning electrically neutral black hole which is as follows:

$$ds^2 = -(1 - \frac{R_s}{r})c^2 dt^2 + \frac{1}{1 - \frac{R_s}{r}} dr^2 + r^2(d\theta^2 + \sin^2(\phi)d\phi^2) \qquad (15.1)$$

where,

$$R_s = \frac{2GM}{c^2} \qquad (15.2)$$

and G is the gravitational constant, M is the mass of the black hole and c is the speed of light. r is the distance of any point from the centre of the black hole. We, as stated, are a long ways from the black hole and in relatively flat space-time. We can see that the Lorentz factor is:

$$\gamma(r) = \sqrt{1 - \frac{R_s}{r}}. \qquad (15.3)$$

So if we look at time dilation on the surface of a black hole, it is the time dilation which we see from a distance. Sometimes people get confused in thinking that time stops on the surface of a black hole and the brick never gets to cross it. But this is only as it appears to an observer from afar. As we all know, what you perceive is not necessarily, (nor sufficiently), reality. So, to the rescue, we bring into play our virtual observer who may be of any race and any age she so desires at any time. Kamis can do this.

The virtual observer shrinks herself to a very small size relative to the size of the black hole and sits on her chair on the surface of the black hole. She is not permitted to cross the surface of the black hole or she will completely leave our consideration forever. Because she is very small all of the spacetime around her is flat. Local spacetime is flat spacetime. Remember the key to relativity is that you are never moving relative to yourself. In General Relativity, that translates to the statement that local spacetime is flat spacetime. The virtual observer's clock is running, according to her, at a normal rate of speed, all her rulers are nice and straight and nothing is amiss. Tidal forces, or spaghettification, has no effect because she is small enough that the curvature of space and time is unnoticeable. She will see the brick cross the event horizon of the black

---

[2]She is a Kami within the Shinto religion.

134

hole just as though there was no event horizon there at all. However, the brick will disappear from her view. This is the difference between a local virtual observer and a distant observer. The metric indicates how a distant observer will measure events in a highly curved space. The local clocks behave, to the local observer, just as clocks that are local to a distant observer also behave. Both are in flat spacetime according to their local observers. However, comparing the two clocks in their different environments results in the two clocks behaving very differently from each other. The brick crosses the event horizon in a finite time in it's own reference frame travelling along with it, or with a virtual observer sitting atop the brick screaming, "Yahoo!" and riding to her certain doom.

The crux is, we can forget the Schwartzchild metric for the purpose of looking at the surface of the black hole in the reference frame at the surface of the black hole itself by a virtual observer positioned on that surface and moving with the reference frame.

Now, consider the surface of a stationary non-spinning black hole, i.e. the event horizon, as a spherical manifold. This manifold has a constant curvature of K. Let us also consider this manifold has radius ρ and a two-dimensional surface described by angles θ and ϕ. We consider the application of a Minkowski space of $(ict, x, y, z) = X_\mu$ where:

$$
\begin{aligned}
ict &= ict \\
x &= R\cos(\theta)\sin(\phi) \\
y &= R\sin(\theta)\sin(\phi) \\
z &= R\cos(\phi)
\end{aligned}
\tag{15.4}
$$

If we now spin the black hole and if γ is the Lorentz transformation of this manifold as a result of the spinning of the black hole, and ds is considered the measure between events, then the metric would be:

$$
ds^2 = \frac{cdt^2}{\gamma^2} - dR^2 - \gamma^2 R^2 \sin^2(\phi)d\theta^2 - R^2\cos^2(\phi)d\phi^2
\tag{15.5}
$$

where

$$
\gamma = \sqrt{1 + \frac{\omega_0^2 R_s^2}{c^2}}
\tag{15.6}
$$

and ω_0 is the angular velocity of the equator of the black hole. Of course γ is a function of R_s which is the distance to the axis of spin of the black hole from any point on the surface of the black hole. The variable R is the distance to the origin of our coordinate system, which we have set at the centre of the black hole. On the surface of the black hole, R is a constant, (this means $R_s = R$),

135

and $dR = 0$. We are only observing what is happening from changes in θ and ϕ.

The manifold determined by the surface of the black hole is the projection of a four dimensional Minkowski space onto a spherical surface. Let us consider the parametric equation of some function $f(t, R, \theta, \phi)$ which is bounded by the dictates of the Lorentz transformations. Note that the function $f(t, R, \theta, \phi)$ is a vector where t will be considered constant for each calculation [3], and R is the Schwartzchild radius which will also be considered a constant. In this way the four dimensional Minkowski space is collapsed and projected onto the surface of a sphere. The angle brackets below denote the dot product of the vectors produced by indicated partial differentiations.

We are using the parameterization of the sphere of radius R centred on the origin as:

$$f(\theta, \phi) = (t, R\sin(\theta)\sin(\phi), R\sin(\theta)\sin(\phi), R\cos(\phi)) \qquad (15.7)$$

where $g_{\mu\nu}$ is the metric tensor describing the geometry of the local region of space-time. The components of the metric tensors are then:

$$
\begin{aligned}
g_{00}(t, \theta, \phi) &= <\tfrac{\partial f}{\partial t}, \tfrac{\partial f}{\partial t}> \\[4pt]
g_{01}(t, \theta, \phi) &= <\tfrac{\partial f}{\partial t}, \tfrac{\partial f}{\partial \theta}> \\[4pt]
g_{02}(t, \theta, \phi) &= <\tfrac{\partial f}{\partial t}, \tfrac{\partial f}{\partial \phi}> \\[4pt]
g_{10}(t, \theta, \phi) &= <\tfrac{\partial f}{\partial \theta}, \tfrac{\partial f}{\partial t}> \\[4pt]
g_{11}(t, \theta, \phi) &= <\tfrac{\partial f}{\partial \theta}, \tfrac{\partial f}{\partial \theta}> \\[4pt]
g_{12}(t, \theta, \phi) &= <\tfrac{\partial f}{\partial \theta}, \tfrac{\partial f}{\partial \phi}> \\[4pt]
g_{20}(t, \theta, \phi) &= <\tfrac{\partial f}{\partial \phi}, \tfrac{\partial f}{\partial t}> \\[4pt]
g_{21}(t, \theta, \phi) &= <\tfrac{\partial f}{\partial \phi}, \tfrac{\partial f}{\partial \theta}> \\[4pt]
g_{22}(t, \theta, \phi) &= <\tfrac{\partial f}{\partial \phi}, \tfrac{\partial f}{\partial \phi}>
\end{aligned}
\qquad (15.8)
$$

The matrix form is then

---

[3] In other words we freeze time and do a mess of calculations at each time step. The black hole will be spinning at some speed for awhile, then we increase the spin and then see what happens.

$$G_{\mu\nu} = \begin{pmatrix} g_{00} & g_{01} & g_{02} \\ g_{10} & g_{11} & g_{12} \\ g_{20} & g_{21} & g_{22} \end{pmatrix} \tag{15.9}$$

We now have a tensor equation which describes the geometry of the surface of the sphere. If we wish to draw curves or examine various lines and paths on the surface of this sphere, we can create a function $\theta(\phi)$, for example, then substitute it into Equation 15.7 to obtain a curve resulting from $f(t, R, \theta(\phi), \phi)$, where R is a constant. We can then derive values required for $G_{\mu\nu}$ and from

$$G_{\mu\nu} = 8\pi T_{\mu\nu} \tag{15.10}$$

obtain the complete values of stresses, strains and shears on the surface of the black hole.

If we draw a sphere with lines of latitude and longitude, we can outline or emphasize some lines of longitude, from the poles to the equator which represent longitudinal lines of which conserve the metric while the black hole, (or coordinate system), is spinning. See Figure 15.1.

It may be helpful go through the logistics of drawing such coordinate lines which conserve the metric. Recalling ancient techniques of mathematics which preceded the use of algebra and was restricted to using only compass and straight edge, without numbers and without coordinate system, the construction of various curves and ability to do things such as bisect angles, construct parallel lines and so on; we may entertain the idea that solutions to mathematical problems during those times relied on the ability to successfully complete geometrical constructions. The Einstein Equations of Equation 8.2 then show the interactions between physical and material interactions as a result of local

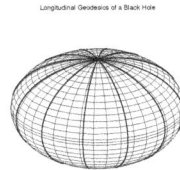

Figure 15.1: Lines of a coordinate system, (longitude), from the poles of a stationary non-spinning black hole to the equator.

geometries. If we are to draw a line from the pole of a non-spinning black hole to the equator, we would simply draw a line of longitude. First, we set R to unity. We then set θ to some value θ_0 and keep it constant while increasing the value of ϕ from zero to $\pi/2$. That line describes the line conserving the metric from pole to equator. On the bottom half, we again keep θ as θ_0 and decrease the value of ϕ from π to $\pi/2$. We would then have a set of values of ϕ while θ and R are constant. We take these sets of values and substitute into Equations

15.4. From there, should one wish, the elements of the Einstein tensor can be determined and resultant stresses and so on be found along the paths of the resulting curved lines of longitude. This is the technique that was used to draw Figure 15.1.

Let us now see what these longitudinal lines would look like if the black hole was spinning. As we traverse, in the local reference frame on the surface of the black hole following the path from the pole to the equator, we would rotate as we moved toward the equator. If some point, $P(t, R, \theta, \phi)$ over which we traverse, was moving with a tangential velocity of v_t then it would have moved through some angle θ_t by the time we had travelled from the pole to the point under consideration. From the faster tangential velocity the local clocks and rulers at the traversing point P changes as it traverses from pole to equator.

$$\gamma = \frac{1}{\sqrt{1 - v_t^2}} \tag{15.11}$$

where v_t is in terms of the speed of light and unit-less. Therefore, if the tangential speed of the surface of the sphere is dependent on its longitude and the measure of time is dependent on the Lorentz factor, then the measure of θ in the local frame of the surface of the black hole is:

$$
\begin{aligned}
v_t &= v_e \sin(\phi) \\
\gamma &= \frac{1}{\sqrt{1 - v_t^2}} \\
\theta &= \gamma v_t
\end{aligned}
\tag{15.12}
$$

where v_e is the tangential velocity of the surface of the black hole at the equator. We show various lines of longitude of black holes spinning with different speeds in Figure 15.2.

Now we toss in a brick.

Let us suppose a brick crosses the event horizon, enters the black hole and leaves our consideration forever. It has added angular momentum to the black hole. Let us also surmise that the brick has no charge and does not add to the electrostatic potential of the black hole. Consider that it has entered the black hole at some point, say at a point of entry of $\pi/6$ radians above the equator. Due to the fact that the brick crosses the event horizon at the speed of light in the local reference frame of the surface of the black hole, a perturbation forms on the function describing the tangential velocity at various points on the surface of the black hole going from the pole to the equator. If the black hole is spinning at maximum spin, then the tangential velocity of the surface is zero at the poles and is the speed of light at the equator. This is described by a smooth and analytical sine function. However, at the entry point, this function forms a soliton. This soliton is an increase in the curvature of spacetime on the surface of the black hole. It can be calculated from the amount of

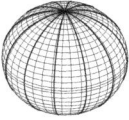

(a) Lines from the poles of a stationary non-spinning black hole to the equator.

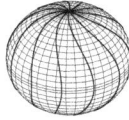

(b) The black hole is spinning an equator moving at half the speed of light.

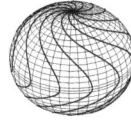

(c) The black hole is spinning with an equator moving at 0.9 times the speed of light.

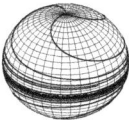

(d) The black hole is spinning with an equator moving at the speed of light.

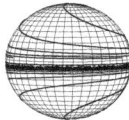

(e) The black hole is spinning with an equator moving at the speed of light. Side view.

Figure 15.2: A progression of black holes spinning at faster and faster rates. Note the bending of coordinate lines of longitude which conserve the metric at the equator until, at the speed of light, it is an infinite path which joins the equator at infinity.

angular momentum added to the surface of the black hole. There is a tidal effect which forms another soliton in the opposing hemisphere. This addition to the overall angular momentum of the black hole would lower the temperature of its surface to below absolute zero unless this additional curvature can be removed from the system. Since there are differences between the stresses below and above the point of entry the perturbations are forced towards the poles. At the poles, the tangential velocity is zero and differentials in curvature

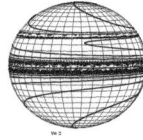

Figure 15.3: Lines of longitude from the poles of a spinning black hole to the equator at the time a brick is thrown into it at a point $\pi/6$ radians above the equator.

along the lines of longitude[4] attempt to force the perturbations into singularities. Let us look at this phenomenon step by step. First let us look at the lines of longitude as the brick crosses the event horizon. See Figure 15.3

Of course, we must not forget about tidal effects. Gravity is not just a set of vectors describing a vector field of forces and so on, there are tidal effects to take into account since gravity is a tensor field involving shears as well as stresses and strains. An every day example is given by the behaviour of tides. The Moon's gravitational field pulls the surface of the oceans towards it to form a high tide. But this effect is not just on the side of the Earth which is facing the Moon. The Moon also pulls the Earth towards it and "leaves behind" the ocean on the far side. As a result there is a high tide on the surface of the Earth which faces the Moon and a high tide on the surface of the Earth that faces away from the Moon. There is a squeeze on the Earth's oceans at the poles which causes the surface of the oceans to bulge both towards and away from the Moon. This is the tidal effect of a gravitational field. This must also be brought into play in the case of throwing a brick into a very fast spinning black hole. The black hole's gravitational field pulls the brick across its event horizon and at the same time, the brick's gravitational field pulls the black hole up just as suddenly in the opposite direction. As a result, a perturbation on the bottom half of the black hole forms which equals the perturbation on the top half. Actually the total curvature which results from the perturbations caused by the brick is evenly distributed top and bottom. Half the curvature is in the top hemisphere and half is in the bottom. The total energy content throughout the surface of

---

[4]Please forgive the use of the term "lines of longitude" to describe the curved lines going from pole to pole on the surface of the spinning black hole. Usually lines of longitude are considered as straight and not curved like this. I hope and trust you get the idea.

the black hole resulting from the change in curvature is equal to the mass of the brick times the speed of light squared. One may simply apply the total energy content of the brick:

$$\rho l_\mu l_\nu = \frac{G_{\mu\nu}}{8\pi} \qquad (15.13)$$

Where ρ is the density of the brick and l_μ is the four-velocity of the brick crossing the event horizon. From there, simply integrate through Equations 15.8 to determine the resultant lines of longitude. We have applied a Maxwellian distribution of curvature due to the perturbation caused by the brick with total curvature on the upper hemisphere equal to the curvature on the lower hemisphere. We have shown the various orientations of lines of longitude from the time the brick crosses the black hole, to its resultant tidal effect and eventual transit towards the poles in Figure 15.4.

The resultant solitons are shock waves and move to the poles at the speed of light. But each perturbation cannot be completely squeezed into one point at the pole. A two-dimensional perturbation of curved space-time cannot be compressed completely into the point where there is no tangential velocity because space-time cannot be curved infinitely. There is a limit to its bending and squeezing into a singular point at the pole, because of the boundary conditions of Schrödinger's Equation. A small "sphere"[5] of space-time forms at the pole. All of the curvature contained in each soliton cannot remain completely on the surface of the black hole. It forms a "pimple" at the pole. A very small part of space-time is lifted slightly from the Schwartzchild radial distance at which the escape velocity is the speed of light and is in a neighbourhood where the escape velocity is less than the speed of light[6]. This bit of space-time is squeezed and forced directly away from the pole at a speed very close to the speed of light. Let us examine this bubble of space-time.

Bringing back Schrödinger's Equation as before... (if you read the previous chapter on Schrödinger's Equation then you don't have to wade through the following and can take a bit of a break while the slow ones get to go through this.)

---

[5]A mathematical ball.
[6]Quantum Mechanics trumps General Relativity

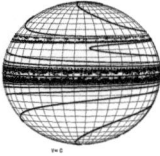
(a) Brick crosses event horizon.

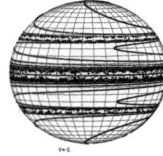
(b) Tidal effects cause a second perturbation.

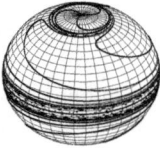
(c) The perturbations are pushed to the poles.

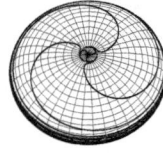
(d) Very close to the poles, the perturbations are squeezed into a mathematical ball determined by the boundary conditions of Schrödinger's Equation.

Figure 15.4: A progression of a perturbation resulting from throwing a brick into a very fast spinning black hole. The initial perturbation at $pi/6$ radians above the equator causes a second matching perturbation below the equator. These perturbations are forced to the poles due to the differences in tension above and below the perturbation caused by differing curvatures on the surface of the black hole.

15.2   Boundary Conditions of Schrödinger's Equation

Consider any particle with mass m and charge resulting in an electrical potential energy V which obeys the following:

$$i\hbar\frac{\partial}{\partial t}\psi = -\frac{\hbar^2}{2m}\nabla^2\psi + V\psi \qquad (15.14)$$

which is derived from:

$$-\hbar^2\nabla^2\psi = \mathbf{p}^2\psi \qquad (15.15)$$

where $\mathbf{p}$ is the momentum of the particle. It is well known that if there are infinite or semi-infinite boundary conditions, the above equation is commonly solved through a singularity solution. We reject assuming infinite boundary conditions and present a well established and common method, known as the separation of variables, to find a general solution in a finite domain following which the boundary condition becomes obvious.

Rewriting the differential equation we have:

$$i\hbar\frac{\partial}{\partial t}\psi = -\frac{\hbar^2}{2m}\left\{\frac{\partial^2}{\partial x^2} + \frac{\partial^2}{\partial y^2} + \frac{\partial^2}{\partial z^2}\right\}\psi + V\psi \qquad (15.16)$$

Let:

$$\psi = T(t)X(x)Y(y)Z(z) \qquad (15.17)$$

where, $T(t)$ is a function of t only, $X(x)$ is a function of x only, $Y(y)$ is a function of y only and $Z(z)$ is a function of z only. Then:

$$i\hbar\frac{\partial}{\partial t}TXYZ = \begin{aligned}&-\frac{\hbar^2}{2m}\left\{\frac{\partial^2}{\partial x^2}TXYZ + \frac{\partial^2}{\partial y^2}TXYZ + \frac{\partial^2}{\partial z^2}TXYZ\right\}\\&+VTXYZ\end{aligned}$$

$$(15.18)$$

Under the condition that $\psi \neq 0$ we can divide through by $TXYZ$ to yield:

$$i\hbar\frac{T'}{T} = -\frac{\hbar^2}{2m}\frac{X''}{X} - \frac{\hbar^2}{2m}\frac{Y''}{Y} - \frac{\hbar^2}{2m}\frac{Z''}{Z} + V \qquad (15.19)$$

We can see that each term is linearly independent. Since each term is being varied by its independent variable and all variables are linearly independent from each other, and the constant term is also independent from the others, each term must equal a constant. Therefore:

$$i\hbar\frac{T'}{T} = -\alpha^2 \tag{15.20}$$

$$\frac{\hbar^2}{2m}\frac{X''}{X} = V - \beta^2 \tag{15.21}$$

$$\frac{\hbar^2}{2m}\frac{Y''}{Y} = -\gamma^2 \tag{15.22}$$

$$\frac{\hbar^2}{2m}\frac{Z''}{Z} = -\xi^2 \tag{15.23}$$

where α, β, γ and ξ are constants, and the equation has been separated. We have placed the constant term, $-V$, with equation 6.7 since it has been chosen as the direction of travel of the particle.

Then:

$$X = \cos\left(\frac{\sqrt{2m(\beta^2 - V)}}{\hbar}x\right) \tag{15.24}$$

We consider a slight re-write as:

$$X = \cos\left(\frac{2\pi\sqrt{2m(\beta^2 - V)}}{\hbar}\frac{x}{2\pi}\right) \tag{15.25}$$

Consider the boundary condition of $X = 1$ which occurs when:

$$x = \frac{2\pi\hbar}{\sqrt{2m(\beta^2 - V)}} \tag{15.26}$$

and

$$x\sqrt{2m(\beta^2 - V)} = h \tag{15.27}$$

since

$$\mathbf{p}^2\psi = -\hbar^2\frac{\partial^2\psi}{\partial x^2} \tag{15.28}$$

and using boundary condition ...

$$\psi = 1 \tag{15.29}$$

we get

$$\mathbf{p}^2 = -\hbar^2\frac{\partial^2\psi}{\partial x^2} \tag{15.30}$$

We also have

$$-\frac{\hbar^2}{2m}\frac{\partial^2 \psi}{\partial x^2} = \beta^2 - V \tag{15.31}$$

yielding

$$-\hbar^2 \frac{\partial^2 \psi}{\partial x^2} = 2m(\beta^2 - V) \tag{15.32}$$

Substitution yields:

$$x\mathbf{p} = h \tag{15.33}$$

at the boundary of the particle. However the "angle" within the cosine goes from 0 to 2π and therefore we have a measure of Δx. Because x varies between the boundaries we have a non-constant $\mathbf{p}$. We therefore have:

$$\Delta x \Delta \mathbf{p} = h. \tag{15.34}$$

Note that any boundary condition other than $\psi = 1$ substituted into equation 15.28 invalidates the previous equation.

We would like to mention here that the boundary would yield a "probability" of one for the particle should ψ represent probability. Inside this boundary this probability would be less than one. At the "centre" of the particle, the probability would be -1 and this is absurd. In the derivation of a solution we had said $\psi \neq 0$. So we will deny the particle to exist inside the boundary and, for that matter, outside the boundary as well. For this particular solution to stand, the particle only exists where $\psi = 1$ and does not exist otherwise. We are stating that the particle does not exist when $\psi < 1$. This is a different case than determining the position or time of the particle. In this case we are determining the existence of the particle itself. We are postulating that if ψ is less than one, then it isn't. We conclude ψ cannot be a measure of probability. It is a potential. When the potential is 1, the particle exists. From these considerations, the particle can only exist at it's boundary.

From outside the particle we can only measure to an accuracy of:

$$\Delta x \Delta \mathbf{p} \geq h \tag{15.35}$$

Considering the temporal characteristic function, we have an exponential of $i\frac{\alpha^2}{\hbar}t$. Let us now consider α. We note the units of measure. We see that $\hbar$ is in units of joules-sec. We see that t is in seconds and will cancel the time unit of $\hbar$ leaving joules in the denominator. Hence, since the exponential must be unitless, α^2 is in units of joules. To continue the discussion allow α^2 to be some unknown form of energy in joules. We will examine what this means as follows.

Let

$$E = \alpha^2 \tag{15.36}$$

so the exponential within the temporal characteristic function becomes

$$\frac{iEt}{\hbar} \tag{15.37}$$

and we look at the situation where $\psi = 1$. In other words, the particle definitely exists. We have seen that at the boundary, from before, the spacial characteristic function is one. Therefore the temporal characteristic function is also one for a temporal boundary. This can only occur should the exponent of the temporal characteristic function be $2\pi i$. In which case we have:

$$\frac{iEt}{\hbar} = 2\pi i \tag{15.38}$$

rearranging

$$Et = 2\pi\hbar \tag{15.39}$$

$$Et = h \tag{15.40}$$

Here we have time going from 0 to some cyclic value yielding an exponent of $2\pi i$. We will then denote this as Δt and ΔE is the magnitude of fluctuation of energy. We now have:

$$\Delta E \Delta t = h \tag{15.41}$$

and observing from outside the particle in the time dimension, we can only measure to an accuracy of:

$$\Delta E \Delta t \geq h \tag{15.42}$$

This happens outside some time ordered "boundary" where/when the potential of the existence of the particle yields $\psi = 1$. Combining both time and spacial ordered factors we have the situation where any measurement of the time, location, momentum or energy of the particle must obey the following;

$$\Delta x \Delta \mathbf{p} \geq h \tag{15.43}$$

and

$$\Delta t \Delta E \geq h \tag{15.44}$$

because that is determined by the boundary conditions of any particle adhering to Schrödinger's equation. Since this has been validated by an overwhelming amount of experimental and, now, theoretical evidence, we propose that the Heisenberg Uncertainty Postulate be classified as a theory.

Let us take a closer look at E.

The exponent of the temporal characteristic function is a phase angle that allows the particle to have a potential of existence equal to one at each cycle.

Let:

$$\frac{E}{\hbar}t = \theta \qquad (15.45)$$

And we differentiate by t on each side to yield:

$$\frac{E}{\hbar} = \frac{d\theta}{dt} \qquad (15.46)$$

or

$$\frac{E}{\hbar} = \omega \qquad (15.47)$$

$$E = \hbar\omega \qquad (15.48)$$

and,

$$E = h\nu \qquad (15.49)$$

So, this energy, E, is not a form of energy coming from the mass of the particle or it's momentum of motion or even it's charge generating V. It appears to be an energy that is associated with the time ordered frequency of the particle's existence. This energy is not associated with mass or charge.

Let us examine α further.

$$\alpha = \frac{\sqrt{2\pi i\hbar}}{\sqrt{t}} \qquad (15.50)$$

and

$$\alpha = \frac{1}{\sqrt{2}}\left(\sqrt{\frac{h}{t}}\right)(1+i) \qquad (15.51)$$

Continuing, we see that we can also say:

$$\theta_n = n^2 2\pi i, \; n \in \mathbb{N} \qquad (15.52)$$

whenever $\psi = 1$. So this exponential has been quantized by n^2. This can be compared to an orthogonal set of eigenfunctions yielding a complete solution of ψ. There are interesting consequences to the general solution of Schrödinger's equation. We call α an eigenvalue in an eigenspace which we often use to find general solutions. Apparently α^2 is the energy of a photon. We are proposing that the magnitudes of an infinite number of eigenvalues to the general solution of Schrödinger's equation yield the energy values of an infinite number of subatomic particles. The first order temporal eigenvalue yields the energy of a photon.

From the behaviour of this class of differential equations, ψ can be considered as a conserved scalar potential field. Since electromagnetic radiation can be thought of as a moving disturbance within a scalar potential field, and this field is conserved, there is a slight alteration in the surrounding potential field should any disturbance move through it. We believe there is the possibility that a bundle of rapidly fluctuating electromagnetic fields moving at the speed of light, commonly known as a photon, would behave as though it had a very small gravitational field.

A cosine function represents a solution for the value of the spacial dimensions of ψ. The boundary conditions are $\psi = 1$. Therefore, if there is a bubble or sphere of space-time and the value of ψ is one at $x = 0$, then the value of ψ is also one at the other boundary where $x = \frac{p}{h}$. Inside the cosine function if θ is the function of x which solves the Schrödinger Equation, then:

$$\psi = cos(\theta(x)), \ \theta = 0, 2\pi$$

and when $\theta(0) = 0$, $\psi = 1$ and when $\theta(p/h) = 2\pi$, $\psi = 1$. However, at the centre $\theta(x = p/(2h)) \to \theta = \pi$ and $\psi = -1$. What can it mean that $\psi = -1$? We speculate that there exists a sphere consisting of the potential of existence. The middle of the sphere, up to half the radius, has a negative potential of existence and the outer shell, past half the radius, has a positive potential of existence. Consider that this potential is realized in the hostile environment just above the pole of the spinning black hole. The positive shell has 7/8 of the volume, while the negative inner sphere has 1/8 of the volume, of this bubble of space-time. There would then be a 1 : 7 ratio between the two volumes. If we postulate that the negative volume ends up being anti-matter and the positive volume ends up being matter, then we would predict a 1 : 7 ratio between anti-matter and matter or 0.14 : 1. However, if the bubble became matter, or condensed into matter and anti-matter, the two types of matter would annihilate each other leaving only 3/4 of the original material. This would result in 1/4 of the material being annihilated in matter-anti-matter explosions. This would result in a great deal of energy. If there is such a fast spinning black hole at the centre of a galaxy, it would have material, probably pure Hydrogen, Helium and some neutrons, pouring out from its poles at a very high speed and have a great deal of energy, probably in the form of gamma and cosmic rays, also being emitted.

In terms of energy and the Stress-Energy tensor, using the Legrangian, we can say [Einstein et al (1923)]:

$$T_{\mu\nu} = \rho l_\mu l_\nu \tag{15.53}$$

where ρ is the energy density of the area we are examining. l_μ and l_ν are the four-velocities of the area.

The curvature of an arc is given as:

$$k = \frac{1}{R} \tag{15.54}$$

where

$$R = G\frac{M}{c^2} \tag{15.55}$$

$$k = \frac{c^2}{GM} \tag{15.56}$$

For a non-spinning black hole, (here T is the temperature):

$$
\begin{aligned}
T &= c^2 \frac{dM}{dA} \frac{1}{C_A} \\[1em]
&= c^2 \frac{dM}{dR} \frac{dR}{dA} \frac{1}{C_A} \\[1em]
&= c^2 \frac{c^2}{G} \frac{1}{\frac{dA}{dR}} \frac{1}{C_A} \\[1em]
&= \frac{c^4}{G} \frac{1}{\frac{d(4\pi R^2)}{dR}} \frac{1}{C_A} \\[1em]
&= \frac{c^4}{G} \frac{1}{8\pi R} \frac{1}{C_A} \\[1em]
&= \frac{c^4}{8\pi G} \frac{1}{R} \frac{1}{C_A}
\end{aligned}
\tag{15.57}
$$

From this we see that either $C_A = 1$ in which case $T = \frac{1}{8\pi R}$, or we may set $C_A = \frac{1}{8\pi}$ which would make the temperature equal to $1/R$ if both c and G are set equal to one. This would have some elegance since it would mean that the temperature would be equal to the curvature of the surface of space-time. In either case, the curvature of a non-spinning black hole is a non-zero constant, so long as R is constant. As the coordinate lines which conserve the metric from poles to equator become longer and longer as the black hole increases its spin, the measure of the curvature of the surface drops to zero. And as a result, the temperature correspondingly drops to zero as well. Under this condition, and only under this condition, is there a possibility for a reversal in the flow of entropy.

Looking at this thermodynamically. We began with a brick. The brick has a lot of information attached to it and has a high amount of entropy. We throw the brick into this fast spinning black hole and we end up with pure Hydrogen,

Helium and radiation. Material coming out has very little information and low entropy. We can make stars out of it. We can't make stars out of bricks. It appears this model would allow for high entropic material to accrete into a black hole and have low entropic material come out of it, along with some energy. Some anti-matter would escape. This is just what we see coming from very fast spinning galaxies having jets of material. The faster the galaxy is spinning, the more prominent the jets.

Furthermore, the jets are shooting straight out from the poles of the spinning black hole. In order for something to escape a gravitational field, it must exit the field at an angle. It cannot escape from exiting the field straight up. The Hydrogen and Helium gasses escaping from the poles of the spinning black hole, would then fall back towards the plane of the galaxy. A very fast spinning black hole would eject renewed star-making stuff vertically from the galactic disk. This stuff would in turn fall back onto the disk as a sprinkler system of Hydrogen and Helium gas renewing the material in the galactic disk. This stuff becomes stars, planets and human beings, which all eventually accrete back into the black hole at the centre of the galaxy, to be lost forever, while the remnant angular momentum so generated causes space-time pimples to burst forth from the poles of the black hole. In this way a galaxy "renews" itself. If the universe is infinite and eternal, there must be some way that there can be a renewal of material out of which to make stars and human beings, or else the entire universe would burn itself out. It ain't pretty, but it works.

15.3   Email from Cam

I got this email from Cam:

> They pop.
> Small, falling, black holes with low angular momentum are formed under the heavy gravitational field of the collapsing galaxy. They bounce off and when they are far enough away they pop, to-day's globular clusters are the popcorn black holes that were too far away to be swallowed by the explosion. This puts their ages within a few billion years of each other as no globular cluster could be older than the life of the galaxy plus one return trip of a bouncing black hole.
> Only popcorn black holes that had the same properties to bounce far enough away would survive. Which is why they are all the same size, and rather large.
> Sent from my iPhone

I have no idea what this means, but it sounds fascinating and worth a look And Werner liked it too.

15.4   The Information Paradox

Definition: The Event Horizon is a location in space where the escape velocity is the speed of light.

Consider a brick – a brick made in Medicine Hat, Alberta, Canada in 1957. The brick has mass. We measure its mass in kilograms. However, we could measure the mass of the brick as a length, namely the radius of the brick if the brick was compressed into a black hole. Nevertheless, if the brick was isolated in space, and our virtual observer was at some point distant from the brick, she would have to have some speed in order to escape the gravitational field of the brick. Let us say that this escape velocity is v_{brick}. And, just to make things interesting, let us say we have a black hole not too far away and the virtual observer is sitting outside the event horizon of the black hole and would also calculate that she would need an escape velocity of $v_{blackhole}$, less than the speed of light, to escape the gravitational field of the black hole. Now, if there are three points, the centre of mass of the brick, the centre of mass of the black hole, and the position of the virtual observer, then there exists only one direction which is perpendicular to the plane upon which sit these three locations. And that is the direction used to calculate the escape velocity from the gravitational fields of both the brick and the black hole. The escape velocity is always calculated in the direction perpendicular from the location to the centre of mass of the object from which the virtual observer is trying to escape. If there are two objects, the total escape velocity is the addition of both escape velocities. Or, $v_{escape} = v_{brick} + v_{blackhole}$.

Now consider that we move the brick far, far away and the virtual observer is sitting at the surface of the black hole. The escape velocity for the virtual observer is the speed of light. Then we bring the brick closer to the black hole and the virtual observer realizes she must move further from the black hole in order to be at a place where the escape velocity is the speed of light. Because if she stays where she is, she will have to escape from both the black hole and the brick and the combined escape velocity will be greater than the speed of light and she will be trapped. In other words, as the brick approaches the black hole, the points determining the location of the event horizon change position. The surface of the black hole itself changes shape because of the presence of the brick. Neat.

151

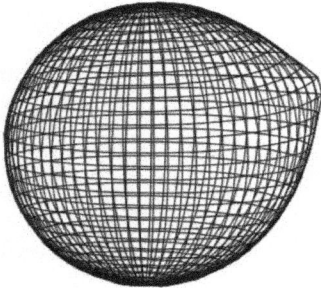

Figure 15.5: The brick alters the escape velocity of its surrounding space and distorts the surface of the black hole.

The brick will cross the event horizon at the speed of light. And, just to make things interesting, the event horizon itself will move outwardly towards the brick at the speed of light. Much like two photons smashing into each other head on. The brick crosses the event horizon of this deformed black hole and leaves our consideration forever. But spacetime acts like an elastic media and the deformed black hole snaps back into the shape of a sphere with a slightly greater Schwartzchild radius. It doesn't wobble radially because the radius of the black hole remains constant. But in the angular directions on the surface of the black hole, there are vibrations of shear, stress and strain. The lines of latitude and longitude vibrate as a result of the brick hitting the surface of the black hole and the surface snapping back into place. The vibrations depend on the brick hitting it and our virtual observer can tell from the vibrations left on the surface of the black hole forever that the vibrations were caused by a brick made in Medicine Hat, Alberta, Canada in 1957.

If the black hole had been spinning at the speed of light as described above, and had been struck by a brick, these vibrations would effect the rate and amount of material that would be ejected at its poles. The information carried by the brick would be ejected back into the universe.

Homework Assignment #4

Consider Pythagoras wherein if there is a right triangle with orthogonal sides x and y with hypotenuse r then:

$$r^2 = x^2 + y^2 \tag{15.58}$$

in two dimensions on a flat surface. If Pythagoras holds for x and y being orthogonal, then the surface on which measurements are taken is flat.

In three dimensions:

$$r^2 = x^2 + y^2 + z^2 \tag{15.59}$$

152

for orthogonal sides x, y and z. If true, then the space is "flat."

Consider two events in a space-time continuum, S_1 and S_2. Then the interval between the two events, if the time interval is dt, and space coordinates are orthogonal, is:

$$ds^2 = -dt^2 + dx^2 + dy^2 + dz^2. \tag{15.60}$$

Should the equation hold true, then we shall denote the space-time region as "flat". We call a region having coordinates, (ict, x, y, z) a Minkowski space. The above equation describes a flat Minkowski space with c, the speed of light, equal to one.

Put this equation into a matrix form. Consider the following "matrix":

$$g_{\mu\nu} = \begin{pmatrix} -1 & 0 & 0 & 0 \\ 0 & 1 & 0 & 0 \\ 0 & 0 & 1 & 0 \\ 0 & 0 & 0 & 1 \end{pmatrix} \tag{15.61}$$

Then we may say:

$$ds^2 = g_{\mu\nu} X^\mu X^\nu \tag{15.62}$$

as a simple matrix equation where:

$$X^\mu = \begin{pmatrix} t \\ x \\ y \\ z \end{pmatrix} \tag{15.63}$$

Now, $g_{\mu\nu}$ has four columns and rows and has two indexes. We call this "matrix" a tensor. $g_{\mu\nu}$ is known as the metric tensor. We present a little homework exercise so the reader may see how it works.

Consider γ such that:

$$\gamma = \sqrt{1 - \frac{2GM}{Rc^2}} \tag{15.64}$$

which is, of course, the red shift factor for light emitted from a massive body with mass M. G is the gravitational constant and c is the speed of light. To change from a flat space or from one coordinate system to another, we simply use $g'_{\mu\nu}$ where:

$$g'_{\mu\nu} = \gamma g_{\mu\nu} \tag{15.65}$$

153

Note, the indexes, μ, ν, are the row and column of the tensor, respectively. This is for a second order, fourth rank tensor. Tensors can have much higher order, but we would need three dimensional paper, which we do not have. The order is the number of coordinates in the space we are considering and the rank is how many indexes we happen to need. In matrix arithmetic, common in most high schools, there is a summation over interior indexes in matrix multiplication. Tensors multiply in the same way however, we do not include the summation sign, we just sum over repeated indexes in any term and just pretend there is a summation sign there. This is known as the Einstein convention. To transpose or change the orientation of a tensor, simply raise or lower the index. Obviously, an upper and lower index represents a dot product. Differentiation is done by using a comma. Differentiate by the coordinate indicated by the index following the comma.

We now save a lot of work by using an operator known as the Chrisoffel Symbol, which is a tensor in its own right, except it has three indexes, and is denoted as:

$$\Gamma^{\mu}{}_{\alpha\beta} = \frac{1}{2} g^{\mu\nu} \left(g_{\nu\alpha,\beta} + g_{\beta\nu,\alpha} - g_{\alpha\beta,\nu} \right) \tag{15.66}$$

We can see here that the index ν is "summed out."

Consider a tensor field S in which the coordinate system is subjected to stresses and strains and consider a path within this tensor field denoted by u having differentiating coordinates λ along such a path. We can now differentiate along u by:

$$\nabla_u S = \frac{\partial s}{\partial \lambda} \tag{15.67}$$

which is comparable to taking the gradient of a scalar potential field.

Then

$$\frac{DS^{\alpha}{}_{\beta\gamma}}{\partial \lambda} = \frac{\partial S^{\alpha}{}_{\beta\gamma}}{\partial \lambda} + S^{\mu}{}_{\beta\gamma} \Gamma^{\alpha}{}_{\mu\delta} u^{\delta} - S^{\alpha}{}_{\mu\gamma} \Gamma^{\mu}{}_{\beta\delta} u^{\delta} - S^{\alpha}{}_{\beta\mu} \Gamma^{\mu}{}_{\gamma\delta} u^{\delta}. \tag{15.68}$$

and when differentiating on a flat space the Christoffel symbols are just 0.

S is a tensor field. The upper indexes correct for curvature, twisting and contraction in the differentiation. and δ is the differentiating index. μ is a dummy index which is interchanged with either α or β and summed for corrections. Add when correcting upper indexes; subtract when correcting for lower indexes. Sum over all repeated indexes.

The assignment is to apply these transformations to the Einstein/Minkowski field equations:

$$G_{\mu\nu} = 8\pi T_{\mu\nu} + 2\left(F_{\mu\alpha}F^{\alpha}{}_{\nu} - \frac{1}{4}g_{m\,u\nu}F^{\alpha\beta}F_{\alpha\beta}\right) \qquad (15.69)$$

along a path bypassing a massive body such as the sun by 1.42 seconds of arc as seen from the Earth.

Don't leave your room until you are done.

Hint: Use the Jacobian on the $(0,0)$ components of the field equations.

Chapter 16

Roxy's Statement

This will actually blow your mind. It did mine.

In Pidgin English – *me belong Roxy*. We have had a little disagreement for the past 35 years. A couple can have rather long-standing discussions and disagreements that last their entire lives. This is actually a good thing. Roxy is a professional tap dancer and I am a mathematician. So we have had to work on this compatibility thing and have had a lot of fun doing so. This little discussion is a disagreement over Truth. I have been saying that Truth is objective and Roxy says it is subjective. Roxy's study is in the arena of social evolution during the late 1700's as well as an in-depth life long study into First Nations cultures. I write poetry.

Nevertheless, Roxy was saying that if we cannot know the difference between reality and our perception then who really cares and that reality is just how we see the world. I am saying that it makes no difference how we see the universe, reality is what is. I am sure you are all familiar with this particular debate. So I continue for 30 years or so on my quest and figure that if ever Roxy agrees with me, then I will truly have something. So I do some math, discover the equation of everything, see Werner a couple of times, discuss and talk about it with Cam, and figure I have it all worked out.

So I'm feeling pretty good here and somewhat proud of myself. And it so happens during springtime in the Yukon that I am leaving a friend's cabin and walking down a mountainside towards town one beautiful sunny morning. The snow is melting, the Sun is up and I am lost in contemplation walking through the jack pine. I looked at the Sun glistening over the mountains in the morning sky and I visualized the light coming from the Sun being bent through extreme curvatures of space and time, perhaps at the centre of our galaxy, and matter being formed. Just like this rock here beside me ... woah! Something flew quickly through the edge of my field of view while I was looking at this rock. It was a mosquito. I was startled and looked around. There, again beside me by the rock, was an intricate spider's web glistening in the sunlight streaming in through the trees. It was breathtakingly intricate and beautiful. Each dewdrop was distinct and each fibre of the web reflected light like a beaded string of rainbow coloured diamonds. This had been made by a living thing. The mosquito was also alive. I stepped back under the weight of this revelation. My God, it is all alive! I looked around at the jack pine, carefully. The pine needles were not just mineral existence, they were living matter. All of the trees were alive, I listened to the forest; it was filled with the sound of birds and humming insects. The deep dark green of the nearby pines and gray-black stands of tree trunks contrasted with shining white melting snow banks; a stunning portrait. This planet, this small segment of the Universe, was teaming with life. How the hell was I going to explain life in the Unified Field Theory? Like, I know that biology is only for kids that can't do physics; but this time the bio students had the upper hand. I also know that I can traverse a line of logic and rational thought all the way from a measure of distance and time to the complex world of chemistry and crystallography, and I know you can add sparks and electrical stimulation to chemicals and get amino acids and I know you can make RNA in a test tube, (I've seen it done); but I also know you can shake, rattle and roll a test tube of chemicals all you like and wait forever and a day and you will never, ever, ever, make a mosquito and have it fly out of that test tube as a result of your efforts. Life is a barrier in the line of logic from mathematics to outward reality. Life – biology – is a river we can only cross on a bridge of Faith. OK, we're alive. This is life. How the hell did we get here?

Needless to say, I was devastated. I no longer felt pride in any of my mathematical or physics accomplishments. It was all for nothing. There was no way I could complete the theory. The universe had been on my side throughout this derivation, but life itself had delivered the fatal coup. I didn't give up; but I knew I was defeated.

Then, a few months later, I looked at dark matter.

Actually, I looked at a digital photo of NGC 3198. I had transformed the pixels of a digital photo of NGC 3198 so that the galaxy would be portrayed as seen from directly above. Then I used a negative of the transformed photo, black on white. I printed this digital photo out. It was obviously a spiral. I knew that. I also knew it was a straight line. I took the picture into the living room and sat on the couch by Roxy. I showed her the photo.

"Hi," I said, "You'll never guess what this is."

158

"Looks like a galaxy," she said.

"Yes," I said, "But I know something that no one else in the world knows. This galaxy looks like a spiral, but it is actually a straight line."

"What do you mean it is a straight line?" she asked.

"The galaxy is rotating," I said, "It is very large. Gravity travels at the same speed as the speed of light. By the time the gravitational influence has gone from the centre of the galaxy to the outer parts, the galaxy rotates and ends up looking like a spiral. But it is really a straight line."

"I don't see that at all," she said, "How can a spiral be a straight line?"

"Remember in New Zealand?" I asked, "Remember when we were far away from everywhere in Taumarunui on the rugby field in the dark at night?"

"Yes," she said.

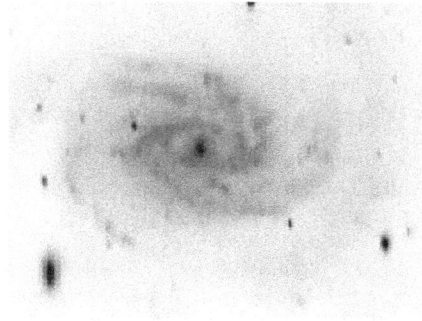

Figure 16.1: The transformed photo of NGC 3198. The galaxy that saved our marriage.

"Remember the centre of the galaxy and what it looked like? Remember the optical illusion, that it is really just a cylinder filled with stars and goes straight into the centre of the galaxy and out the other side. That it all is just a straight cylinder of stars?" I said.

Roxy looked confused.

I continued, "This is just a spiral. If I can figure out the equation of the spiral it will be one of the greatest discoveries in all of science. It will completely change our understanding of the universe and physics itself."

"Can you do that?" she asked.

"Yes," I said, "but it will take a little time."

"How much time?" she asked.

"Couple of weeks," I said.

"Well what the hell are you doing sitting here on the couch?" she demanded, "Get back in there and get the formula!"

So I did.

159

And Cam, my son, got very excited and we had this formula and I was busy measuring the distances to galaxies and it was all very much fun, except for Roxy.

I walked into my office one morning to find Roxy staring at the computer displaying the transformed photo of NGC 3198. She was in tears.

"I don't get it," she said, "You and Cam say it all so simple and I don't get it. How can this be a straight line? How is it a spiral?"

I didn't know what to say or do.

"Don't try so hard," I ventured, "It is not so difficult, it is subtle. Just try to let the ideas come to you rather than forcing yourself to understand it. It's OK."

Se was still very frustrated and I walked quietly out of my office leaving her to her frustration. I didn't know what to do and she was feeling hurt. I was sad. The next day I was cleaning up the kitchen when Roxy walked in and asked:

"Do you mean to tell me, that we have all these really intelligent and important scientists who are studying astronomy, looking through telescopes, and they study this stuff. You mean to tell me ... I mean ... are they looking at the same galaxy I have looked at and they haven't taken into account that the stars they are looking at have moved by the time they are seeing them?"

"I don't know," I said, "If I ever get the chance, I'll ask them."

"You've got to be kidding," she said, "Even I know that. Those stars are thousands of light years away. I knew that as a child. They haven't figured out that the stars are no longer where they are looking at them? I don't believe that."

And she stormed out of the kitchen.

The next day, Roxy came up to me in the hallway.

"I get it!" she said forcefully, "I get it. I get what you have been talking about all these years. The mathematics isn't the result of the physics and what we are looking at; the physics is the result of the mathematics! It's so obvious.

"This bookshelf," she exclaimed striking the bookshelf beside me with the palm of her hand, "This bookshelf is a bookshelf because it follows the mathematics of being a bookshelf. We don't come along and see the bookshelf and then make up a mathematics of a bookshelf and then find another bookshelf and have to make up an entirely new mathematics to describe it. The mathematics says that this is a bookshelf. The mathematics determines the physics. The physics does not determine the mathematics. I get it!"

I was totally awed by her passion; she had discovered the pinnacle of mathematical and theoretical physics. "You have now reached the place where I am,"

I said, "That is the height of my mathematical insight and I cannot go any further. If you stick to it, and remember what you have just discovered, no one can argue scientifically against you. You have the key to it all."

She looked confident with an air of anger at the world of academia.

I then had to tell her about biology and how my theories were defeated by the existence of life. I told her about being in the Yukon and that I could not complete the theory. However, a great deal of it seems to be covered, but there was no way I could include biology or life in the theory of everything.

A few days later Roxy and I went for an evening walk. It was one of those rare days in Calgary when it is not snowing and it is reasonably warm. There were blossoms on the trees. It was nice. After walking for a while Roxy started talking:

"I notice," she began, "That there is order in the universe. If we look at the smallest things like atoms, and protons and electrons there are rules that tell us what they do, how they behave. We have quarks and they have rules on how they behave. And even smaller than that, what makes up quarks and so on and so on. Things get infinitely smaller and smaller without end. And each world, smaller and smaller, no matter how much we search, there will always be rules that govern them.

"There is always order," she continued, "Nothing is chaotic; nothing is random. There are laws that govern the world of the quantum, the world of the atoms. And there are rules of chemistry. It is not chaotic. There is order in chemistry and in physics. And if we look at a blade of grass," she said stopping on the sidewalk to point to a lawn beside us, "There is order describing the structure of a blade of grass and even how it grows. It may be very complicated and we may not understand it, but that doesn't matter. There is order in the way a blade of grass grows because that is what makes it a blade of grass.

"Also," she went on pointing out a bough of apple blossoms just overhead, "There is order in the growth of a tree, in its leaves and branches and its roots. It is not just growing randomly, there is order to it.

"And the Sun and the planets," she said as we looked up at the evening sky, "There is order in how the Earth rotates and goes around the Sun, in the behaviour of earthquakes and the weather and the ecology. We may not understand it, but there is still order to it all. Nothing is chaotic. Nothing is random. The way the solar system is, the planets going around the Sun, the way the Sun and stars go around the galaxy; there is order to it. It is all ordered. And how all the galaxies are, and the clusters and super clusters of galaxies however they are and even bigger than that, to an infinitely large structure of never ending structures containing structures. It is worlds within worlds, infinitely large and infinitesimally small, and throughout it all there is order. Nothing is chaotic.

"And," she said stopping to face me, "If you are going to make a bookshelf, or a building, you have to have some plan in order to build it. If you just take mud and sticks and try to make a building without some plan, the building will not stand; it will have no structure. It will collapse. It won't be a building. If there is order to everything, then there has to be a plan; there has to be a blueprint for anything to exist for it to have order.

"Therefore," she said to me, "Since there is order throughout the universe, the universe has to have a blueprint. And the blueprint of the universe is mathematics.

"And you, my Love, have discovered a very small piece of the blueprint. This thing with galaxies, it is just a small part of the blueprint. And because you have uncovered some small part of the blueprint, you can see, not completely and only a very small bit, but nevertheless, you can see how everything works. That's why you can see it and no one else can. You have discovered a small part of the blueprint of the universe."

I have been speechless ever since. The universe has a blueprint and the blueprint of the universe is mathematics. The universe has order and nothing is chaotic. In viewing reality, probability does not determine outcomes.

And so, we have a response to the Copenhagen Interpretation. And I, nearing the end of my life, am so humbled and grateful to have married Roxy.

Chapter 17

The Universe is Infinite and Eternal

The title of this chapter says it all. An objective approach to reality leads us to no different conclusion. The universe is not static. It is evolving and we are a part of that evolution. As we live our lives, we disturb the space and time around us. And that disturbance is forever. And it not only affects eternity, it affects the infinite, just as the infinite and eternal affects us.

The universe, the space between the atoms, empty space, is evolving in a particular direction. It is as though there is a flow to the universe through time. There is some direction or path in which things are moving. We are a part of that motion. We have talked about goodness and truth. Are these qualities contained within the coordinate system describing space and time? From what we have seen here, the answer is yes. If you can determine the existence of truth, and the fact that the universe constructs everything out of truth, it is not a great leap to verify the existence of goodness and love. There has been so much ridicule regarding love, and yet we cannot live without it.

We have also looked at the ultimate decision in life: to choose whether or not to be of service to others, and that love is a by-product of a life of service. There is a relationship with others that is part of the makeup of the universe. There is, what I would refer to as, homogeneous and heterogeneous knowledge. Homogeneous knowledge is a very simple type of knowledge which is very limited. It contains truth, but truth is stifled and has no interaction with "the other". Heterogeneous interactions grow and evolve. There are constant interactions which make more knowledge. It is the interaction between the one and the other that causes change – that causes things to happen. For example, we have presented here an argument that truth is objective, that reality is ... real. But that is an argument that stops short. Today there is a denial of the objectivity of truth. And if truth is considered to be only subjective, then it leads to madness. If truth is seen to be only objective, it cannot grow and becomes void of passion. There would be no passion, no love. It is in the interaction of the subjective, of one's perceptions, with the reality of one's surroundings, that gives life to mere existence.

We have perception, which is never completely accurate, and is constantly changed by circumstance and experience. This is called learning. And learning is a very real thing. Something I have recently learned from spending two years with the Blackfoot, (they are so far ahead of Eurocentric epistemology), when our experience has brought us to a certain point, when we are ready, when we are in tune with the many relationships contained within our surroundings, then, from the void, knowledge is allowed to form in our minds from our experiences and previous learning. This knowledge forms when building blocks of experience are in place and, more importantly, when it is deemed that we are ready and fit, by some external standard. In a way the knowledge is given to us but it is also created within us. The source of knowledge lies with the individual. This is the result of a heterogeneous interaction between the objective and the subjective.

As far as we know, we are completely unique in the universe. We are also essential to the universe in order to manifest the information created in the past; and this information can only be manifested through a heterogeneous interaction. In other words, between people. We may have experiences with nature, but our experiences with people are orders of magnitude more powerful. And this experience with others is not geared to be successful by controlling others but by submitting to that which is best for others. It is how we evolved. It is how we are.

17.1   Free Will

Yes, we have free will. Much of the denial of the existence of free will comes from Laplace's Paradox, which basically says that if everything is controlled by deterministic equations, which they are, then once everything is known from some starting point, then all else stemming from that starting point is predetermined by these equations. That would be true if there was a starting point. But there is no starting point. The eternal has no beginning. And it has no end. Nothing is predetermined. There was no starting point. There cannot be a predetermination. There are physical laws and there are mathematical laws governing those physical laws. The outcomes are not predetermined because there cannot be a completely determined initial condition. We can get really good at predicting results from what we know and experiments will result in repeated outcomes given repeated inputs. However, there is biology.

We can even get really good at training animals and people. For example, we can train a flatworm, which will swim away from light as a result of millions of years of evolution, to go against its instincts and swim towards light by giving it electric shocks if it tries to swim away. But there is always that stubborn flatworm that refuses to swim towards the light. Electric shock punishments don't work for everyone. Furthermore, our refusal to succumb to things like addictions goes against the predicted outcomes of psychology. Only human beings can cure themselves of addictions; an animal can not. Sometimes it is the rebelliousness in us that displays our ability to decide and sometimes it is our decision to submit to the discipline of some regimen, like mathematics. It depends. We are really complicated and really difficult to understand. There is no mathematical formula to determine human behaviour. There is a lot of knowledge on, generally, how people will react to certain events, but we are not machines. In a way, there is outer space, which is infinite and eternal, but there is also inner space, within our minds, within our thoughts, which has infinite capacity and may have its beginnings in a past eternity.

Chapter 18

What Does it all Mean?

This has been an interesting journey. We have come to a plateau and may be able to draw a few conclusions.

Firstly, ψ is the potential of existence. We also see that Truth itself is the essence of existence.

We see that the space-time continuum is the infinite array of a four dimensional coordinate system. The coordinate system is a measure between events. This measure involves numbers. Numbers exist. The coordinate system, therefore, exists.

A simplified coordinate system could be compared to a ruler. The ruler measures distance. Whether there are numbers printed on the ruler or not is irrelevant; they exist. There are a couple of rulers, half the ruler, etc. The numbers exist as a measure provided by the ruler. The ruler may be straight or we could perhaps bend it. But what is the ruler made of?

The ruler, the lines of the space-time continuum, the coordinate system at the root of the Einstein/Minkowski equations, is made of ψ. The potential of existence, as determined by Schrödinger, ψ, forms the coordinate system which is the universe itself. The rulers, the coordinate system made out of ψ, can be bent or "curved". We propose that the second time ordered differential of ψ, and the biharmonic of ψ, are measures of this curvature. The more the coordinate system is bent or curved, the greater the potential of existence. This coordinate system can only be bent so far. The limit of bending is determined by Heisenberg. At that limit, the lines of the coordinate system, the rulers, crimp and, in effect, become "knotted" into a particle. The energy of creating the particle from the bending of space-time is equal to its mass times the speed of light squared.

But what is ψ?

Very simply, we conclude that ψ is the potential of existence, and existence has the potential to be verified, to be true. As we develop and pursue Truth, we raise the potential of Truth until this potential achieves reality. If the universe is flat, there is no potential of existence and nothing exists. Nothing can exist, not ever.

This is a time of great excitement within the field of gravitational study. We postulate that types of information form different complexities of Truth, which continue to exist within the vibrating stresses and strains on an event horizon as determined by Schwartzchild.

Furthermore, there is nothing to guarantee that the laws of physics have remained constant throughout eternity. They may well have changed. Perhaps, to paraphrase and alter Einstein's famous quote, God is not only subtle, He has a particularly nasty streak when it comes to pushing back the frontiers of science.

If ψ is the potential of existence and truth is existence itself and the foundation of virtue, perhaps it would not be outside the realm of reason to conjecture that the entire universe somehow consists of "Goodness". The question I would like to ask the reader is: Who in their right mind would go and create and entire universe out of Goodness?

Truth is objective, not subjective. Truth is. Truth is in and of itself. Truth is what it is. Consider the following:

Truth is beauty, beauty Truth.
That is all ye know in the world,
And that is all ye need to know.

- John Keats

There have been a number of hindrances to discovering a Unified Field Theory. One is the inability to derive Heisenberg from Schrödinger. However, we see that this inability is merely the result of pursuing an incorrect philosophy based on an incorrect assumption. It is remarkably easy to derive Heisenberg from Schrödinger given a more reasonable assumption at the beginning. I would also like personally to add that Heisenberg used h rather than $\hbar$ in his uncertainty equations. I have fought long and hard to show that Heisenberg was right in the first place and people should not have messed with his formulae. He remarked quite often that people have misinterpreted his theory to mean a great deal more that it was supposed to.

Einstein had worked for the latter half of his life to try to find the Unified Field Theory. Many have said he was not that good at mathematics. They are wrong. Einstein was a brilliant mathematician. Of course, you are free to go through his papers and point out any errors in his calculations. So far, I have been unable to find any myself. But don't let that stop you.

The axiomatic approach to discovering mathematical truths has been used for thousands of years and has been assailed throughout. In the early 1800s there was a movement to overthrow the axioms in order to substantiate arbitrary measures in the dissemination of justice. For the past hundred years or so it has succeeded and as a result humanity has suffered through many wars and abject poverty disguised under the cloak of materialism. The world is now burdened by those in authority who are doing everything possible to control everyone and they can't even control themselves.

There are alternative societies to ours. And these are societies that have existed for thousands of years. There is no reason for us to cling to a way of life based on the premise that we must all increasingly consume stuff made from non-renewable resources or else we will all die. We have ended up being forced to work ardently to obtain more and more stuff we simply don't need. Life is not that hard if we work together. The things humanity needs in this day and age are not material; they are of a higher nature. They are spiritual and relate to the human spirit. Only people can recognize Truth and its inherent beauty.

I am nearing the latter part of my life. It has been a good life having met and worked with incredible and wonderful people. Over the years I have seen that most people are good and follow a path of integrity. Sooner or later we all come to a point of decision, Adlerian, [URL-14], I guess, where we can choose to suffer the consequences of integrity, or get away with lying. It may not even be that big a deal. We can lie and no one will really know, no one of consequence anyway. Or we can tell the truth and even though it is no great deal, suffer for it. We have all had to come to that point of decision at some time in our lives. However we decide, determines the rest of our lives. Falsehood multiplies. Decisions come again and again and the point of falsehood that may have set us on that path must be justified, rationalized and covered up. People who live a lie go mad. Of course, the opposite is true too. If a decision is made to maintain integrity, even though one may suffer for it, that decision just seems to keep coming back in different forms. And the honest cannot get ahead and the dishonest always appear to succeed. And our society is built for the successful. So it seems that on the whole, if one chooses a life of integrity, one has chosen a life of poverty and hardship. This is not always the case, but it is more often true than not.

So there you go. Truth is and so are you. You can follow it or not according to your desire. And no one says you have to play the fool. You can live a lie or you can acknowledge that the best things in life are free and live in service to others rather than living a life trying to get others to be of service to you. However, you cannot rationalise your decision. Whatever it is – you wear it.

Homework Assignment #5

He aha te mea nui?

Chapter 19

Answers to Selected Problems

Assignment #1

2) About 10^{-7}

4) About 9 billion light years

6) About 55 Kilometres per second per Megaparsec

8) About $3°$ K

Assignment #2

This is a very fun equation to play with. This should help. First we "do" the harmonic equation, which also solves the biharmonic, as follows:

First we resolve the harmonic equation, which also solves the biharmonic, as follows:

$$\psi^*\left(t, r, \theta, \phi\right) = T\left(t\right) R\left(r\right) \Theta\left(\theta\right) \Phi\left(\phi\right),$$
$$\frac{d^2}{dt^2} T\left(t\right) = -\alpha^2 T\left(t\right),$$
$$\frac{d^2}{dr^2} R\left(r\right) = -\alpha^2 R\left(r\right) + \beta^2 \frac{R(r)}{r^2} - 2\frac{\frac{d}{dr}R(r)}{r},$$
$$\frac{d^2}{d\theta^2}\Theta\left(\theta\right) = -\Theta\left(\theta\right)\beta^2 + \frac{\Theta(\theta)\gamma}{(\sin(\theta))^2} - \frac{\cos(\theta)\frac{d}{d\theta}\Theta(\theta)}{\sin(\theta)},$$
$$\frac{d^2}{d\phi^2}\Phi\left(\phi\right) = -\gamma^2\Phi\left(\phi\right)$$

Where:

$$\frac{\partial^2}{\partial t^2}\psi^*(t,r,\theta,\phi) = \\ \begin{pmatrix} 2r\sin(\theta)\frac{\partial}{\partial r}\psi^*(t,r,\theta,\phi) \\ +r^2\sin(\theta)\frac{\partial^2}{\partial r^2}\psi^*(t,r,\theta,\phi) \\ +\cos(\theta)\frac{\partial}{\partial\theta}\psi^*(t,r,\theta,\phi) \\ +\sin(\theta)\frac{\partial^2}{\partial\theta^2}\psi^*(t,r,\theta,\phi) \\ +\frac{\frac{\partial^2}{\partial\phi^2}\psi^*(t,r,\theta,\phi)}{\sin(\theta)} \end{pmatrix}\frac{1}{r^2(\sin(\theta))} \qquad (19.1)$$

and

$$\begin{aligned} T(t) =\ & A\sin(\alpha t) + B\cos(\alpha t) \\ R(r) =\ & \frac{C}{\sqrt{r}}\,BesselJ\left(1/2\sqrt{1+4\beta^2},\alpha r\right) \\ & +\frac{D}{\sqrt{r}}\,BesselY\left(1/2\sqrt{1+4\beta^2},\alpha r\right) \\ \Theta(\theta) =\ & E\,LegendreP\left(1/2\sqrt{1+4\beta^2}-1/2,\sqrt{\gamma},\cos(\theta)\right) \\ & +F\,LegendreQ\left(1/2\sqrt{1+4\beta^2}-1/2,\sqrt{\gamma},\cos(\theta)\right) \\ \Phi(\phi) =\ & G\sin(\gamma\phi) + H\cos(\gamma\phi) \end{aligned} \qquad (19.2)$$

Now we resolve the linear part which is:

$$\frac{\partial^2\psi(t,r,\theta,\phi)}{\partial t^2} = k\psi(t,r,\theta,\phi) \qquad (19.3)$$

having solution:

$$\psi(t,r,\theta,\phi) = f_1(r,\theta,\phi)e^{\sqrt{k}t} + f_2(r,\theta,\phi)e^{-\sqrt{k}t} \qquad (19.4)$$

Now, we will show a little of the biharmonic part in Cartesian coordinates.

Note that if:

$$\frac{\partial^4\psi(x)}{\partial x^4} = \alpha^4\psi(x)$$

$$(D^4 - \alpha^4)\psi(x) = 0$$

$$(D - \alpha)(D + \alpha)(D^2 + \alpha^2)\psi(x) = 0 \qquad (19.5)$$

$$(D - \alpha)(D + \alpha)(D - i\alpha)(D + i\alpha)\psi(x) = 0$$
$$\psi(x) = Ae^{\alpha x} + Be^{-\alpha x} + Ce^{i\alpha x} + De^{-i\alpha x}$$

$$\psi(x) = F\cos(\alpha x) + G\sin(\alpha x) + H\cosh(\alpha x) + I\sinh(\alpha x)$$

So, both the real and the imaginary parts to the harmonic equation work as a solution to the biharmonic equation in Cartesian coordinates. In this case, we can use both x and ix in the harmonic term for a solution to the biharmonic term which doubles the number of solutions for each dimension; however, only half can be used at a time.

Some explanation here. Suppose we have a Fourier Solution to some problem. We can see that we can have both a cosine series and a sine series as solutions. It may be tempting to use both the sines and cosines in a mixed series to try to solve for undetermined coefficients. I know, I was tempted too. But you cannot do that because the general equations, both real and imaginary, (cosine and sine), are repeats of each other. Both are equivalent. So you can't mix them. The real solution, the cosine one, is perfectly valid and may be chosen because of boundary conditions. Or the imaginary solution, the sine one, is also perfectly valid and may be chosen instead because of boundary conditions or initial conditions or whatever. But you cannot chose both of them. You have to choose one of the other. Because of this, if you have 20 solutions in Eigenspace, you can only use either real or imaginary parts of the solutions in any one dimension because in any one dimension the real and imaginary solutions are identical and over determine the overall solution. Usually it is preferable to reject either the real or imaginary parts because of boundary conditions. In the x direction, say, you may possibly choose a cosine solution and reject the sine solution. And in the y direction, to continue the example, you may choose the imaginary solution and reject the real solution. So you cannot use all 20 eigenvectors; you can only use 10 of them. However, you are in a 20 dimensional space but you can only detect 10 dimensions for any event. Then for the next event, you may only detect 10 out of the twenty eigenvectors but they may be a different set of eigenvectors from the previous event. Trippy.

You can then apply the boundary conditions to the equation in spherical coordinates to solve for the arbitrary constants resulting in the appearance of zeta functions.

It may be interpreted that in eigenspace each of the eigenvalues, α, β, etc. are summed over an infinite number of values in such a way that each term of the solution is orthogonal in order to match the boundary conditions. In Cartesian coordinates we could have α_n, β_n, γ_n and ξ_n and possibly sum n^2 over the general solution as n goes from one to infinity. Each value of n results in an "eigenset" and each set of eigenvalues forms a vector space containing orthogonal eigenvectors. These eigenvalues therefore form a multidimensional orthogonal space. There are twelve spacial for the biharmonic term, six for the harmonic term and two for the temporal term. That makes a 20-dimensional eigenspace. However, only half the eigenspace can be used at a time as previously explained. We therefore have a ten-dimensional eigenspace.

So, both the real and the imaginary parts to the harmonic equation work as a solution to the biharmonic equation. You should be fine from here. You can then apply the boundary conditions to the equation to solve for the arbitrary constants[1].

Assignment #3

1) $T = \frac{1}{m}$ in relativistic-quantum silly units[2]. We have to use the units of no units, (sigh), that means we use units in which the speed of light, Plank's constant and the gravitational constant are all one.

2) No.

3) See Kerr and the Kerr metric, [Kerr (1963)], [URL-17].

4) Absolute zero which occurs when the angular momentum equals the mass of the black hole squared. Black holes are very cold to start with, about 10^{-20} degrees K.

Assignment #4

A factor of about 10^{-7}.

Assignment #5

He tangata. He tangata. He tangata.

---

[1] If this equation ever works out to have any physical meaning whatsoever, I will be far more surprised than anyone.

[2] Werner called them silly. Who am I to argue?

Acknowledgements

The author gratefully wishes to acknowledge assistance from the following people:

- Cameron Rout, Yale University

- Dr. Werner Israel, University of Victoria

- Dr. Keith Promislow, Michigan State University

- Dr. Teddy Schmidt

- Dr. Gary Margrave, University of Calgary

- Fred Babbot, Rothney Observatory, Priddis, Alberta, Canada.

Appendices

Appendix A

Description of L-1 Space

What is L-1 space Bruce? What do you mean it is non-Euclidean? Please explain.

L-1 space is a mathematical space in which you are restricted to a single axis. To be a little clearer, in four-dimensional space we can move about in three dimensions. The ability to move is the result of the existence of time, the fourth dimension. Or zeroth dimension. Two dimensional space is a space of only two dimensions like a surface. But we usually consider dealing with the surface of a flat piece of paper. Please read Flatland: A Romance of Many Dimensions, [Abbott, Edwin A. (1884)] immediately. In the book Flatland, Abbot considers life living on a flat surface. He describes another place called Lineland in which creatures are restricted to living on a straight line with no other dimension, just length. What Abbot has so famously described, is a space known as L-1 space. However, the lines do not have to be straight if seen from a higher-dimensional space. But in L-1 space, everything appears to be on a straight line, the line you inhabit. This is a very different place than higher dimensional spaces. This, mathematically, is a very dangerous place. In this space you are not allowed to to a lot of things you can do in higher dimensional spaces. And if you forget that, you end up coming to conclusions that simply don't exist. Like Dark Matter.

From the work of Lebesgue and Minkowski, [Lebesgue (1902)], [Minkowski (1910)], we define a norm as:

$$L_p = \left[\sum_{i=1}^{N} ||X_i||^p \right]^{1/p} \tag{A.1}$$

where p denotes the degree of the measure $||X_i||$ although $||X_i||$ is usually called a norm but you are perfectly ok to think of it as a measure in the usual way. Like the distance between two points for example. If we have $p = 2$ then we are in L-2 space and the measure is an L-2 norm. Note that N is the dimension of the space. If we let $N = 3$ in L-2 space we have the measure calculated, if you expand equation (A.1) as:

$$L_2 = \sqrt{x^2 + y^2 + z^2} \tag{A.2}$$

which is the tried and true Pythagorean theorem in three dimensional space. In other words the Pythagorean theorem is just an L-2 norm in whatever multidimensional space you like. This is known as L-2 space. It is a very safe space to work in. Note that the p-norm of the Lebesgue measure is not the same as the dimension you are working in. It is a statement of the type of measure you are making. Where ever you can make a measure, that is known as a Banach space. Please do not confuse this with solutions of the diffusion equation, which is not a Banach space. An L-1 norm does not apply to solutions of the diffusion equation and therefore does not apply in the world of Quantum Mechanics. Let us move on.

Let us now look at $p = 1$ in equation (A.1), as did Minkowski. Let us say our measure is ds. We would usually say $ds = \sqrt{dx^2 + dy^2 + dz^2}$ however, that is a trap. We cannot go there. In L-1 space, we have to say $ds = dx + dy + dz$. There are severe consequences to this. Let us now look at Euclid.

Five axioms of Euclid are:

1. All things equal to the same thing are equal to each other. (definition of equality)

2. Equals added to equals make equals. (definition of the uniqueness property of operations)

3. All right angles are the same. (orthogonality exists)

4. Given two lines and a transecting line crossing them, the two lines shall cross on the side of the transecting line which contains the sum of interior angles which is less that two right angles. (definition of parallelism and the existence of direction)

5. The whole is equal to the sum of its parts. (definition of conservation)

And for the sake of simplicity, if all the above points are met, you are in a Euclidean space. L-2 space is a Euclidean space, ie. $p > 1$. L-4 space is a Euclidean space. However, looking at L-1 space, you are restricted to staying on a line. the only measure you have is length. So lengths can equal each other. And there can be equal lengths. But there is no direction. There is only staying on the line and no deviation from it. So we cannot have orthogonality or direction. But lengths are conserved. So, applying the Euclidean axioms to L-1 space, we have:

1. All things equal to the same thing are equal to each other. (definition of equality)

2. Equals added to equals make equals. (definition of the uniqueness property of operations)

3. deleted

4. deleted

5. The whole is equal to the sum of its parts. (definition of conservation)

And almost all of the stuff we use in mathematics cannot be used. All we can do is add and subtract lengths. There can be no differentiation, for example. There can be mass, but it has to be all the same along the line. No forces, no velocity, no momentum. In short, no freaking Gravity. Nada. And that is where everybody got into a lot of trouble trying to figure out how a galaxy works. I find that so unbelievably cool.

A.1   Galaxies and L-1 Space

In order to transform into L-1 space, let us project material along the spiral arms onto a straight radial axis wherein the material and direction of forces are aligned along the path of a geodesic in a non-rotating coordinate system. In effect, each star along the spiral arm is influenced by the combined gravitational shadow of other stars and behaves according to its local set of clocks and rulers. Note that gravitational influences in a rotating system follow the path of a geodesic which would appear as a straight line in both rotating and non-rotating systems. Also note that in L-1 space, there is no orthogonality or direction other than along the radial axis. Other than tidal effects, all forces, which are in the radial direction, add up to zero and without reference to material outside of the system, the orientation appears locally static.

A Lebesgue, [Lebesgue (1902)], measure considers n coordinates of some space having an order of p. The p-norm or L_p-norm of x is defined by:

$$L_p = \left(\sum_{i=1}^{n} ||x_i||^p \right)^{(1/p)}. \qquad (A.3)$$

The Euclidean norm, where $p = 2$, is referred to as L-2 space. When $p = 1$, this is referred to as the norm of an L-1 space, [Minkowski (1910)].

L-1 space is a mathematical space in which one is restricted to movements along a single axis and in which the measure is absolutely convergent. In a 4-dimensional Minkowski space, the measure of gravitational forces is the sum of inverse squares and is therefore absolutely convergent. Furthermore, the path of least time is restricted to a single axis. To obtain a linear axis from L-2 space, an orthogonal curvilinear coordinate system is used in which one coordinate is a set of Archimedes' spiral arms and it's orthogonal counterpart is a set of hyperbolic spirals.

To expand further on this point: Consider the Earth-Moon system. Looking outwardly from the surface of the Earth to the Moon, we see that the same face of the Moon is always pointed towards the Earth. The period of rotation of the Moon is equal to its revolutionary period. We see the Moon orbit the Earth each month. However, from the surface of the Moon, things appear very differently. The Earth has a stationary position in the Moon's sky. The sun, stars, planets and so on appear to revolve about the Moon once per month, but the Earth itself does not. It appears as a stationary object above the surface of the Moon. Because the Earth and Moon revolve about a common center of mass with slightly elliptical orbits, the Earth, as seen from the Moon, appears to approach the Moon slightly and then recede once per month. There is a precession of the plane of their mutual orbits which would also cause the Earth to wobble as it spun in the Moon's sky with a period of wobble of about 19 years. The Earth, as seen from the surface of the Moon, is stationary in the sky except for some minor oddities.

Furthermore, as seen from the Moon, the Earth is rotating on its axis every 24 hours; however, the oceans on the surface of the Earth appear as a bulge pointing directly at the Moon. Also, consider that there is a mountain range on the Moon which continually points directly at the Earth.

This two body system, as seen from the Moon, is a physical example of a mutually gravitationally bound L-1 system. If all else in the Moon's sky is ignored except for the Earth, no forces appear to be at play other than tidal effects which keep high tides on Earth facing the Moon and the mountain range of the Moon facing the Earth. The two bodies are in a quasi-static linear orientation to each other and the gravitational influences between them follow a geodesic which is a straight line with a very small, almost undetectable, spiral deviation.

A.1.1 Tidal Effects and Stability of a Linear Orientation

The model I presented is that of a linear orientation of material having a constant linear density rotating about a mutual centre of mass which appears as a spiral as a result of relativistic effects. We show here that tidal effects would tend to orient material to match this model. We begin by describing tidal effects in a two body problem using general relativity in order to demonstrate and then consider the stable orientation of n bodies. We first give a brief review of the derivation and then present analytical results.

Consider a Minkowski space having coordinates (ict, x, y, z) where:

$$
\begin{aligned}
ict &= ict \\
x &= r\cos(\theta)\sin(\phi), \\
y &= r\sin(\theta)\sin(\phi), \\
z &= r\cos(\phi),
\end{aligned}
\tag{A.4}
$$

and

$$
r = \sqrt{x^2 + y^2 + z^2}.
\tag{A.5}
$$

The Lorentz factor in a rotating coordinate system is:

$$
\gamma_\omega(r) = \sqrt{1 + \omega_0^2 r^2/c^2}
\tag{A.6}
$$

resulting in the following metric of a rotating coordinate system in the absence of gravitational effects:

$$
ds^2 = \frac{cdt^2}{\gamma_\omega(r)^2} - dr^2 - \gamma_\omega(r)^2 r^2 \sin^2(\phi)d\theta^2 - r^2\cos^2(\phi)d\phi^2
\tag{A.7}
$$

where r is the distance from the center of rotation and θ is the angle from the axis. ω_0 is the angular velocity as measured by a non-rotating observer.

The Lorentz factor within a gravitational field is:

$$
\gamma_s(r) = \frac{1}{\sqrt{1 - 2GM/(rc^2)}}
\tag{A.8}
$$

and the resultant metric is:

$$
ds^2 = \frac{cdt^2}{\gamma_s(r)^2} - \gamma_s(r)^2 dr^2 - r^2\sin^2(\phi)d\theta^2 - r^2\cos^2(\phi)d\phi^2.
\tag{A.9}
$$

183

In order to determine the overall effects of rotation and gravity using general relativity, we shall examine the resultant curvature of their associated accelerations.

Thus it is shown that from the above Lorentz factors,

$$\frac{2GM}{r} \tag{A.10}$$

and

$$\omega^2 r^2 \tag{A.11}$$

can form scalar potential fields in units of velocity squared. To directly combine the effects of two different scalar potential fields, some appropriate way to derive the combined effects of gravity and rotation using general relativity needs to be determined. A difficulty arises since the Lorentz factor of a gravitational field affects radial measures while the Lorentz factor of rotation affects tangential measures. Therefore, we consider equivalent accelerations derived from each potential field taken separately and combine them using vector addition.

Note that the gradient of an overall potential field, ã, resulting from the interaction of these potential fields, would be in units of acceleration.

The principle of equivalence shows that accelerations have identical effects on the local metric regardless of the cause of the accelerations. This presents a means to combine different scalar potential fields. We define these accelerations as follows.

Determining Curvature Using Vector Addition

Consider a scalar gravitational potential field, ψ_s, surrounding a body of mass M. At distance r, let us define this scalar potential field as:

$$\psi_s = \frac{2GM}{r} . \tag{A.12}$$

We shall denote ψ_s as the Schwartzchild potential. The gradient of this field is

$$\nabla \psi_s = -\frac{2GM}{r^2} . \tag{A.13}$$

which has units of acceleration towards the source of the gravitational field. If one were to accelerate at this rate away from mass M, one would observe the path of a beam of light to be bent with the same curvature as the curvature of a geodesic in the gravitational field generated by mass M at a distance r.

Let

$$\tilde{\mathbf{a}}_s = -\frac{2GM}{r^2 c^2} .$$

(A.14)

Note that this can be re-written as

$$\tilde{\mathbf{a}}_s = -\frac{R_s}{r^2}$$

(A.15)

where R_s is the Schwartzchild radius of mass M.

Concerning a rotating reference frame, let

$$\psi_\omega = \omega^2 r^2$$

(A.16)

where r is the distance from the center of rotation and ω is the angular velocity. Then

$$\tilde{\mathbf{a}}_\omega = \nabla \psi_\omega / c^2$$

(A.17)

which forms another vector field consisting of an acceleration.

Then a vector sum can be derived as

$$\tilde{\mathbf{a}} = \tilde{\mathbf{a}}_s + \tilde{\mathbf{a}}_\omega.$$

(A.18)

Vector Field of Curvature in a Two-Body System

The vector addition of gradients applied to rotational and gravitational potential fields described above can be used to determine the stable orientation of matter from tidal effects in a rotating system.

Let us consider two objects having mass m_1 and m_2 in an isolated space in circular orbits at distances r_1 and r_2 about their mutual center of mass with angular velocity ω such that special relativistic effects are negligible.

The gravitational fields of both these bodies combine to effect the measure of overall curvature in nearby space. A vector field resulting from the acceleration due to circular motion can be added to that.

The mutual rotation of these masses is determined by both their masses and their distance from each other; therefore, ω can be stated in terms of r_1, r_2, m_1 and m_2 and substituted into $\tilde{\mathbf{a}}_s$, resulting in the components of the overall vector field. These are

185

$$\tilde{a}_x = \frac{2G}{c^2}$$

$$\left(\frac{m_1(r_1-x)}{\left((r_1-x)^2+y^2\right)^{3/2}} + \frac{m_2(r_2-x)}{\left((r_2-x)^2+y^2\right)^{3/2}} + \frac{m_2 x}{r_1(r_1-r_2)^2} \right)$$

and $\qquad\qquad\qquad\qquad\qquad\qquad\qquad\qquad\qquad\qquad$ (A.19)

$$\tilde{a}_y = \frac{2G}{c^2}y$$

$$\left(\frac{m_1}{\left((r_1-x)^2+y^2\right)^{3/2}} + \frac{m_2}{\left((r_2-x)^2+y^2\right)^{3/2}} + \frac{m_2}{r_1(r_1-r_2)^2} \right).$$

Setting

$$\begin{aligned}
r_1 &= 1, \\
r_2 &= -1, \\
m_1 &= 1, \\
m_2 &= 1, \\
G &= 1 \text{ and} \\
c &= 1;
\end{aligned} \qquad\qquad (A.20)$$

we show the resultant vector field in Figure A.1. This vector field corresponds to the components of an overall curvature tensor and to stresses and strains along the field lines which in turn describes the components of an overall stress tensor. Consequently, the direction of vectors in Figure A.1 indicate the stresses affecting the flow of matter. This trend is towards and along the rotating axis of the two bodies which corresponds to present theory and observations of tidal effects.

Applying this solution to a scenario whereby these two bodies consisted completely of unconsolidated similar particles, the particles would orient themselves along the axial line of a geodesic between centers of masses of the original two bodies. The resultant linear orientation would then be stable.

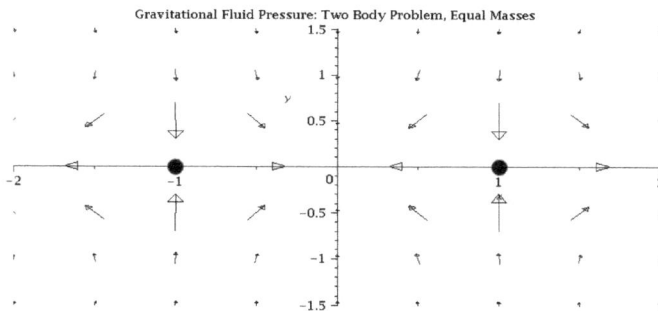

Figure A.1: Vector field of a viscous gravitational fluid surrounding two bodies in mutual circular obit. The two bodies are indicated by dark dots. Note the direction of the "tidal flow" towards the horizontal axis. This diagram was generated using Maple's fieldplot command using equations (A.19) and (A.20). The dark dots at $x = 1$ and $x = -1$ represent the positions of the two bodies discussed in the text.

Appendix B

Radio Map of North Galactic Pole Reveals Influx of Hydrogen

B. Rout, Z. Cutter, T. Fairbrother, K. Pantherbone,
J. Black Plume, M. Stevens, I. Big Tobacco

Siksika Nation High School

B.1   Abstract

A 21-cm radio map around the north galactic pole reveals both an influx of Hydrogen with a velocity of 25-30 kps and and outflux of Hydrogen at about 2 kps. Data in this region shows a distinct double peak in signal strength vs velocity centred at 1.42 GHz. The map shows a quasi-circular region of Hydrogen influx centred on the north galactic pole and having a diameter of about 10 degrees of arc.

B.2   Radio Map

The region near the north galactic pole emits a double peak at the 21 cm line. Part of the emission shows a positive velicity of 2 kps and the other part shows a negative velocity of about -20 kps as in Figure B.1.

Data Aquisition

Data files were downloaded from 49 locations near the north galactic pole. The files were then collated and read for velocity peaks, both positive and negative. The velocities and locations were then used in a half-wave cosine Fourier decomposition in order to produce maps in the region.

Measuring points were at about 5 degrees of arc along right ascension and declination from 12:00 hrs to 14:00 hours and from 10 degrees to 40 degrees. A requested beam width of 0.9 degrees was chosen from reach location to obtain maximum beam width in each measure. Each accessed data point returned the nearest location to that requested, within a few arc minutes, and the accessed location was used for the Fourier decomposition.

B.3   A Galactic Sprinkler System

The double peak is compatible to observing a jet of material from the galactic core which then falls back onto the galactic plain, much like a sprinkler system of Hydrogen gas ejected from the galactic core.

B.3.1   Data Survey

Data was collated from the HI Survey server of the Argelander-Institut fÃ¼r Astronomie. HI All-Sky-Profiles (EBHIS/GASS/LAB) were used.

Effelsberg-Bonn HI Suryey (EBHIS): Effelsberg 100m telescope, $\delta > -4°$
Galactic All Sky HI Survey (GASS): Parkes 64m telescope, $\delta < 1°$
Leiden/Argentine/Bonn HI Survey (LAB): Villa Elisa 30m and Dwingeloo 25m telescopes.

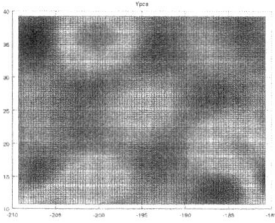

(a) Contour map of receding Hydrogen gas. Centre of map is at the galactic pole.

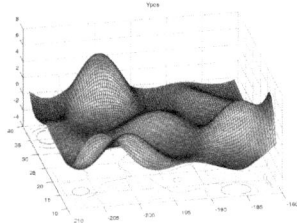

(b) Three dimensional map of approaching Hydrogen gas. Centre of map is at the galactic pole.

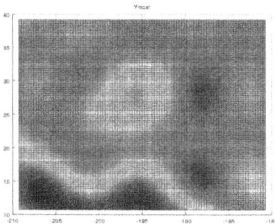

(c) Contour map of approaching Hydrogen gas. Centre of map is at the galactic pole.

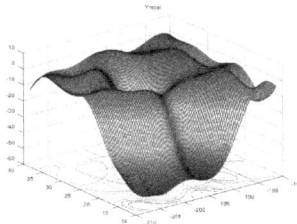

(d) Three dimensional map of approaching Hydrogen gas centred on the galactic pole.

Figure B.1: Hydrogen gas in contour maps showing Hydrogen receding from the north Galactic Pole and approaching the north Galactic Pole. The double peak of the H1 profile shows material receding slowly and then approaching more quickly.

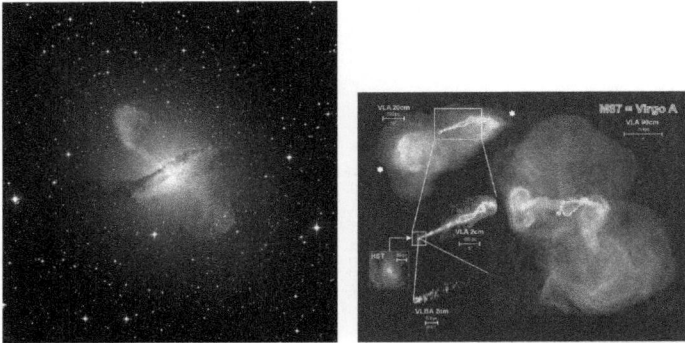

(a) ESO Centaurus A showing distinct jets which fall back onto the galactic plane.

(b) M87 in the HI line showing galactic jets falling back onto the galactic plane.

Figure B.2: Two galaxies, Centaurus A and M87 contain prominent galactic jets which can be seen being ejected from the galactic core and orthogonally to the galactic plane. The ejected material appears to traverse some distance and then fall back onto the galactic stellar region.

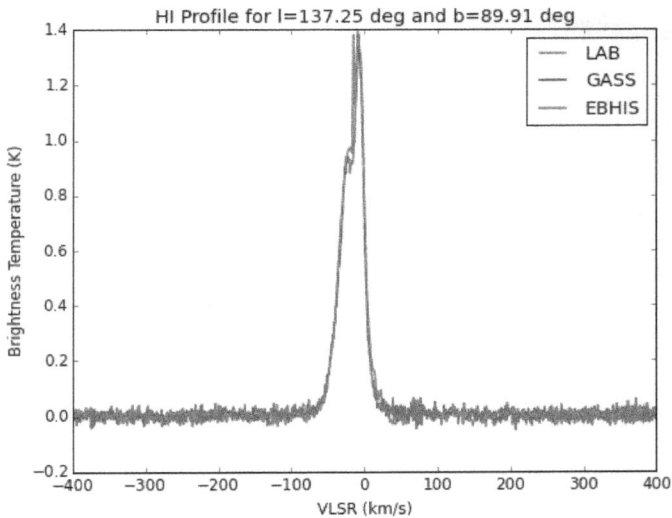

Data Page Link

This interface allows you to extract HI profiles from the EBHIS, GASS, and the LAB survey. The profiles are generated on the fly using a weighted interpolation with a Gaussian kernel. The effective beam size includes telescope beam and beam smearing due to interpolation. The beam for the GASS is 14.4 arcmin (64m Parkes telescope), for EBHIS it is 10.8 arcmin (100m Effelsberg telescope), while for the LAB it is 36 arcmin for declinations > -27.5 deg (25m Dwingeloo telescope) and 30 arcmin for declinations < -27.5 deg (30m Villa Elisa telescope).

Due to interpolations the minimum effective FWHM beam is 16 arcmin for the GASS, 12 armin for EBHIS, and 40 arcmin for the LAB for declinations > -27 deg and 35 arcmin for declinations < -27 deg. This implies for the profiles a different FWHM beam if you select an effective beamsize < 36 arcmin. If you want to compare the calculated column densities, the FWHM needs to be taken into account. Expected uncertainties are in each case at a 2-3% level (about 1% scale uncertainty and 1% for uncertainties in the correction for stray radiation). Additional contributions may be due to noise, residual baseline errors, and RFI (causing occasionally scale errors for the LAB). Column densities are calculated for $-400 < v < 400$ km/s. [1]

---

[1] https://www.astro.uni-bonn.de/hisurvey/AllSky_profiles/index.php

Appendix C

Solving Non-Linear Differential Equations

C.1 The General Operator

Dr. Teddy Schmidt and I were working for the Alberta Research Council in the early 1980s. We lived in Fort Saskatchewan north of Clover Bar in Edmonton and used to commute to work together each morning. We had been working on what we both called "the problem," which is the Stefan Problem in a finite domain, a problem that was 120 years old. I have since solved this problem in a closed form as well as one other known as the diffusion equation with variable diffusion coefficient. I will use these two solutions as examples.

This is the classical Stefan problem in a finite domain:

The diffusion equation:

$$\frac{\partial \psi}{\partial t} = \frac{\partial^2 \psi}{\partial x^2} \tag{C.1}$$

at the left boundary ...

$$\frac{\partial \psi}{\partial x} = 0 \text{ at } x = 0 \tag{C.2}$$

initial conditions ...

$$\psi(0, x) = 0 \tag{C.3}$$

$$R(0) = L \tag{C.4}$$

where L is the length of the media.

Conditions at the right moving boundary are . . .

$$\psi = \psi^* \text{ at } x = R(t) \tag{C.5}$$

$$\alpha \frac{dR}{dt} = \frac{\partial \psi}{\partial x} \text{ at } x = R(t) \tag{C.6}$$

where ψ^* is a constant ≤ 1 and α is also a constant.

Notice that the boundary conditions are coupled with the differential equation describing the interior. To put it simply, if we have a nice solvable differential equation describing the interior of the interactions, the boundary conditions are incredibly complex and deemed unsolvable. If we go through a number of transformations to make the boundary well behaved, the differential equation describing the interior becomes non-linear and is also deemed insolvable. We had been concentrating on this problem for some time since it was a good example of a non-linear problem. Neumann had solved the problem assuming the medium was infinite, but the finite case was still open. The reason we felt this problem was so important to solve in an analytical and closed form was not for the solution itself, but to find out the method by which it could be solved and perhaps such a method could be applied in many other instances.

It was a cold morning in Fort Saskatchewan. The temperature had dropped to minus 40. There is a joke in Alberta that when it's minus 40, it's so cold it really doesn't matter if it's Farenheit or Centigrade. As all Canadians who winter in the north know, the rubber of the tires on the cars freeze solid during the night while they're parked. Usually we think of wheels as being round, but there is a flat part at the bottom of the wheels where the car sits on the road and the wheels are frozen with this flat part on each wheel. When the car is started and drives off, it bumps along the road bouncing with each revolution of the wheels as the frozen flat part hits the ground. Pretty well every Canadian in the north knows about it. So Teddy and I are bouncing along in the car with the heater blasting against the wind screen to keep the front window defrosted while our frozen breath, as a result of our excited conversation, attempted to defeat the efforts of the straining defroster.

"I've been thinking about the problem," said Teddy, "It seems that mathematics is able to deal with linear things and that the world of nature is not linear, it's non-linear. All we can do mathematically when we apply our solutions to nature is make approximations.

"I am thinking," he continued, "that we already have linear things solved. We have, if you follow the idea, a general linear operator. What we need is a general non-linear operator but I find that the number and classifications of non-linear problems is infinite and probably uncountable."

I thought for a bit and said, "No Teddy, we don't need to find a general non-linear operator. what we need is a general operator."

Teddy perked up. "That is a very good idea," he said, "If we can find a general operator, then we subtract the linear operator and the non-linear operator is just what's left behind."

Put into mathematical terms consider the following:

Let R (for Reimann) be the set of all operators. Let $\mathcal{L}$ (for Laplace) be the set of all linear operators and Λ (for Einstein) be the set of all non linear operators. We then have:

$$R = \mathcal{L} + \Lambda \tag{C.7}$$

After working on other approaches for 23 years, I went back to this idea and rearranged the formula.

$$0 = -R + \mathcal{L} + \Lambda \tag{C.8}$$

This describes the problem we have. There are three independant terms in this high level formula which describes many mathematical situations. However, we usually are working in a simplified one dimensional case with an independant variable and the function upon which it depends. In other words we have two variables and three independent terms. This seems to be the crux of the problem. It becomes almost impossible to separate the variables so that the equation can be solved. What is neccessary is to do something, by hook or by crook, to transform the problem in some way; to reduce it to having only two terms. They may be incredibly non-linear, but at least with two terms you have chance of solving the problem.

First Example:

To demonstrate, a perturbation approach was used to the problem with various simple transformations to yield:

$$f'' + \frac{(f')^2}{\alpha} + \frac{z}{2}f' = 0 \tag{C.9}$$

Let

$$v = f' \tag{C.10}$$

Then

$$v' + \frac{v^2}{\alpha} + \frac{z}{2}v = 0 \tag{C.11}$$

This becomes a very nice non-linear differential equation with three terms as described by the general operator approach. We must reduce to two terms. We use the following transformation ...

$$v = e^{-z^2/4} w \tag{C.12}$$

yielding ...

$$v' = (\frac{-z}{2} w + w')e^{-z^2/4} \tag{C.13}$$

we obtain ...

$$(\frac{-z}{2} w + w')e^{-z^2/4} + \frac{w^2}{\alpha} e^{-z^2/2} + \frac{z}{2} w e^{-z^2/4} = 0. \tag{C.14}$$

Rearranging:

$$\left[\frac{z}{2} w e^{-z^2/4} - \frac{z}{2} w e^{-z^2/4} \right] + (w')e^{-z^2/4} + \frac{w^2}{\alpha} e^{-z^2/2} = 0. \tag{C.15}$$

simplifying ...

$$w^2 e^{-z^2/4} + \alpha w' = 0 \tag{C.16}$$

rearranging ...

$$\frac{w'}{w^2} = \frac{-e^{-z^2/4}}{\alpha} \tag{C.17}$$

And we can see the equation has been separated, or uncoupled, and can be solved in a straight forward method from there.

Let us try another example.

Second Example:

Consider:

$$\frac{\partial \psi}{\partial t} = \frac{\partial \psi}{\partial x} \psi \frac{\partial \psi}{\partial x} \tag{C.18}$$

from Fick's law of diffusion with the diffusion coefficient depending linearly on ψ.

This becomes:

$$\frac{\partial \psi}{\partial t} = \left(\frac{\partial \psi}{\partial x} \right)^2 + \psi \frac{\partial^2 \psi}{\partial x^2} \tag{C.19}$$

Here again, we have three independant terms, and must reduce to two terms. We examine:

198

$$w = \psi^2 \tag{C.20}$$

then

$$\frac{\partial w}{\partial t} = 2\psi \frac{\partial \psi}{\partial t} \tag{C.21}$$

$$\frac{\partial w}{\partial x} = 2\psi \frac{\partial \psi}{\partial x} \tag{C.22}$$

and

$$\frac{\partial^2 w}{\partial x^2} = 2 \left(\left(\frac{\partial \psi}{\partial x} \right)^2 + \psi \frac{\partial^2 \psi}{\partial x^2} \right) \tag{C.23}$$

and, of course, we now have what we're looking for. Substituting back we have

$$\frac{\partial w}{\partial t} \frac{1}{2\sqrt{w}} = \frac{\partial^2 w}{\partial x^2} \tag{C.24}$$

There are various approaches from here. But one might be to assume

$$w = \Theta \Phi \tag{C.25}$$

where Θ is a function of t only and Φ is a function of x only.

$$\frac{\Phi \Theta'}{2\sqrt{\Theta}\sqrt{\Phi}} = \Theta \Phi'' \tag{C.26}$$

rearranging yields,

$$\frac{\Theta'}{2\Theta\sqrt{\Theta}} = \sqrt{\Phi}\Phi'' \tag{C.27}$$

and, of course, the problem has been decoupled and a solution can be found. A complete solution involves the Heaviside function involving a number of resultant Γ functions that I do not want to include here, but if you want to satisfy your curiosity, you can try

$$\psi = -\frac{x^2}{6t} \tag{C.28}$$

and see that it works just fine. It's not the compete solution, it just demonstrates that a solution has been found in a closed form. There is mathematics involved in this paper, however it can be reduced to something quite manageable.

To summarize, we have found a handy little device to solve some non-linear differential equations. These two have been presented to demonstrate the technique and to categorically state that just because a non-linear differential equation has been presented, it doesn't mean that a closed form solution can't be found. Above are two classic problems that have been solved fairly easily using this method.

Bibliography

[Abell (1975)] Abell G. O. 1975, Exploration of the Universe, (3rd ed.: Holt, Rinehart and Winston), 621

[Abbott, Edwin A. (1884)] Abbott, Edwin A., Flatland: A Romance in Many Dimensions. New York: Dover Thrift Edition, 1884, (1992 unabridged).

[Afanasev et al (1991)] Afanasev V. L., Zasov, A. V., Popravko G. V., & Silchenko O. K. 1991, SvAL,17, 325A

[Albada et al (1985)] Van Albada T. S., Bahcall J.N., Begeman K. & Sancisi R., 1985, ApJ, 295, 305

[Archimedes (225 BC)] Archimedes, 225 BC, On Spirals, (Encyclopædia Britannica, 2008)

[Aristotle (335 BC)] Aristotle, (circa 335 B.C.) The Basic Works of Aristotle, ed by Richard Mckeon, Random House, Inc., New York, NY 10019 USA.

[Ball (2010)] Rouse Ball W. W., A Short Account of the History of Mathematics, Dover Books on Mathematics, August 19, 2010.

[Barnard (1914)] Barnard, S. and Child, J. M., Elements of Geometry, Parts I - VI, MacMillan & Co. Ltd., St. Martin's Press, New York, 1959. First edition, 1914.

[Begeman (1989)] Begeman, K. G. (1989). *HI Rotation Curves of Spiral Galaxies*. Astronomy and Astrophysics, #223, pp 47-60. Retrieved from the High Energy Astrophysics Division at the Harvard-Smithsonian Center for Astrophysics.

[Binney & Tremain (2008)] Binney, J. & Tremain, S. 2008, Galactic Dynamics, Second ed., (Princeton University Press, Princeton, New Jersey), 456, 480

[Biviano et al (1990)] Biviano, A., Giuricin, G., Mardirossian, F., & Mezzetti, M. 1990, ApJs, 74, 325B

[Braatz et al (1955)] Braatz J. A., Reid M. J., Humphreys E. M. L., Henkel C., Condon J. J. & Lo K. Y. 2010, arXiv:astro-pc/1005.1955v1

[Braatz et al (2009)] Braatz J. A., Reid M. J., Henkel C., Condon J. J. & Lo K. Y. 2009, A White Paper for the Astro2010 Survey, (NRAO MPC Publications)

[Brownstein & Moffat (2006)] Brownstein, J. R., & Moffat, J., W. 2006, ApJ, 636, 721

[Burbidge et al (1963)] E. Margaret Burbidge, G. R. Burbidge and K. H. Prendergast, *The Velocity Field, Rotation and Mass of NGC 4258*, 1963, APj, 138..375B

[Braine et al (1993)] Braine, J., Combes, F. & van Driel, W. 1993, A&A, 280, 451B

[Carter et al (1973)] Carter, B., Hawking, S., Bardeen, J., *The Four Laws of Black Hole Mechanics*, 1973, Commun. math. Phys. 31, 161-170 (1973), Springer-Verlag

[Chemin et al (2003)] Chemin, L., Cayatte, V., Ballkovski, C., et al. 2003, A&A, 405:, 89

[Cooney et al (2012)] Conney, A., Dimitrios, P. & Zaritsky, D. 2012, arXiv:1202.2853v1

[Cooperstock & Tieu (2005)] Cooperstock, F. I. & Tieu, S. 2005, arXiv:astro-ph/0507619v1

[Comte (1978)] Comte, G. 1978 IAUS, 77, 30C

[Crook et al (2007)] Crook A. C., et al. 2007, ApJ, 655, 790C

[Desmonde (1951)] Desmonde, William, H. 1951, J Exp Educ, 19, 3

[Devereaux et al (1992)(] Devereaux, N. A., Kenney, J. D., & Young, J. S. 1992, AJ, 103, 784D

[Einstein et al (1923)] Einstein, Albert, Lortntz, H. A., Weyl, H.. Minkowski, H., The Principle of Relativity, (New York, N.Y.:Dover Publications Inc., 1923)

[Einstein (1905)] Einstein, A., (1905). " Über einen die Erzeugung und Verwandlung des Lichtes betreffenden heuristischen Gesichtspunkt (trans. A Heuristic Model of the Creation and Transformation of Light). Annalen der Physik 17: 132-148.

[Einstein (1914-17)] Einstein, Albert, The Foundation of the General Theory of Relativity, (pp. 146-200 in translation volume), Princeton University Press, 1997, reprinted from The Collected Papers of Albert Einstein, Volume 6, The Berlin Years: Writings, 1914 - 1917, A. J. Kox, Martin J. Klein, and Robert Schulmann, editors.

[Ferrarese et al (2000)] Laura Ferrarese, Holland C. Ford, John Huchra, Robert C. Kennicutt, Jr., Jeremy. Mould, Shoko Sakai, Wendy L. Freedman, Peter B. Stetson, Barry F. Madore, Brad K. Gibson, John A. Graham, Shaun M. Hughes, Garth D. Illingworth, Daniel D. Kelson, Lucas Macri, Kim Sebo and N. A. Silbermann. A Database of Cepheid Distance Moduli and Tip of the Red Giant Branch, Globular Cluster

Luminosity Function, Planetary Nebula Luminosity Function, and Surface Brightness Fluctuation Data Useful for Distance Determinations. The Astrophysical Journal Supplement Series Volume 128, Number 2, Citation: Laura Ferrarese, et al 2000 ApJS 128 431.

[Fish (1961)] Fish, R. A. 1961, ApJ,1 34, 880F

[Freedman (2001)] Freedman, W.L. 2001, ApJ, 553, 47F

[Gallo & Feng (2010)] Gallo, C. F. & Feng, J. Q. 2010, J Cosmol, 6, 1373

[Gil De Pas et al (2007)] Gil De Pas A.,Boissier S., Madore B. F. et al. 2007, ApJs, 173, 185G

[Genzel et al (2008)] S. Gillessen, F. Eisenhauer, S. Trippe, T. Alexander, R. Genzel, F. Martins, T. Ott, *Monitoring Stellar Orbits around the Massive Black Hole in the Calactic Center*, Draft Version, December 10, 2008.

[Halmos (1974)] Halmos, Paul, *Naive set theory*, Princeton, NJ: D. Van Nostrand Company, 1960. Reprinted by Springer-Verlag, New York, 1974. ISBN 0-387-90092-6 (Springer-Verlag edition).

[Herrnstein et al (1999)] Herrnstein J. R., Greenhill L. J., Diamond P. J., et al. 1999, arXiv:astro-ph/9907013v1

[Hubble (1936)] Hubble, E. P. 1936, The Realm of the Nebulae, (New Haven: Yale University Press)

[Jacoby et al (1992)] Jacoby, G. H. et al. *A critical review of selected techniques for measuring extragalactic distances*, PASP 104, 599-662 (1992).

[Jech (2003)] Jech, Thomas, 2003. *Set Theory: The Third Millennium Edition*, Revised and Expanded, Springer. ISBN 3-540-44085-2.

[Józsa (2007)] Józsa, G. I. 2007, A&A, 468, 903

[Kassin & Weiner (2006)] Kassin, S. A., de Jong, R., & Weiner, B., J. 2006, ApJs, 643, 804

[Kepler (1619)] Kepler, J. 1619 The Harmony of the World, (self published)

[Kerr (1963)] Kerr, RP (1963), *Gravitational field of a spinning mass as an example of algebraically special metrics*, Physical Review Letters 11: 237-238. doi:10.1103/PhysRevLett.11.237

[Knapen et al (1993)] Knapen J. H., Beckman J. E., Cepa J., Soledad del Rio, M., Pedlar A., 1993, ApJ, 416, 563

[Kuhn & Nasar (2002)] Kuhn, W. Harold & Nasar, Sylvia, *The Essential John Nash*, Princeton University Press, 2002.

[Leavitt (1908)] Leavitt, Henrietta S. *1777 Variables in the Magellanic Clouds*. Annals of Harvard College Observatory. LX(IV) (1908) 87-110

[Lebesgue (1902)] Lebesgue, H. 1902, *Intégrale, longueur, aire* (Paris: Université de Paris)

[Lindblad et al (1996)] Lindblad, P. A., Lindblad, P. O. & Athanassoula, E. 1996, A&A, 313, 65

[Madore et al (1999)] Madore, B. F. et al. *The Hubble Space Telescope Key Project on the Extragalactic Distance Scale. XV. A Cepheid distance to the Fornax Cluster and its implications*, Astrophys. J. 515, 29-41 (1999).

[Mathewson et al (1992)] D. S. Mathewson, V. L. Ford and M. Buchhorn, *A Southern Sky Survey of the Peculiar Velocities of 1355 Spiral Galaxies*, Astrophysical Journal Supplement Series, 81:413-659, 1992, August.

[Menzies & Mathews (2006)] Menzies, D. & Mathews, G. J. 2006, arXiv:gr-qc/0604092v1

[Minkowski (1910)] Minkowski, H 1910, Geometrie der Zahlen, (Leipzig and Berlin)

[Misner et al (1973)] Misner, Charles W., Thorne, Kip S., Wheeler, John Archibald, *Gravitation*, W. H. Freeman; 2nd Printing edition (September 15, 1973)

[Moore & Gottesman (1998)] Moore, E. M. & Gottesman, S. T. 1998, MN-RAS, 249, 353

[Newton (1687)] Newton. I., PhilosophiæNaturalis Principia Mathematica, 1687, (Cambridge: University of Cambridge)

[Persic & Salucci (1995)] Massimo Persic & Paolo Salucci, *Rotation Curves of 967 Spiral Galaxies*, Astrophysical Journal Supplement Series 99:501-541, 1995, August.

[Pisano et al (1998)] Pisano, D., J., Wilcots, E., M., & Elmegreen, B., G. 1998,AJ,115,975

[Python (1969)] Monty Python's Flying Circus. *The Dead Parrot Sketch*. BBC, 1969 (?)

[Rohlfs & Kreitschmann (1980)] Rohlfs, K. & Kreitschmann, J. 1980, A&A, 87, 175R

[Rubin & Ford (1970)] Rubin, V. C. & Ford, W. K. 1970, ApJ, 159, 379

[Rubin et al (1999)] Rubin, V. C., Waterman, A., H., & Kenney, J. D., P. 1999,ApJ, 118, 236

[Russell (1902)] Russell, Bertrand (1902), *Letter to Frege*, in van Heijenoort, Jean, From Frege to Gödel, Cambridge, Mass.: Harvard University Press, 1967, 124-125.

[Russel (1914-19)] Russel, Bertrand (1914-19), *The Philosophy of Logical Atomism*, reprinted in The Collected Papers of Bertrand Russell, 1914-19, Vol 8., p. 228

[Sartre (1956)] Sartre, Jean-Paul, *Being and Nothingness*, Translated by Hazel E. Barnes, University of Colorado, Washington Square Press, published by Pocket Books, New York, 1956.

[Schopenhauer (1818)] Schopenhauer, Arthur, 1818, *The World as Will and Representation*, Dover. Volume I, ISBN 0-486-21761-2. Volume II, ISBN 0-486-21762-0

   [1] Schrödinger, Erwin (November 1935), *Die gegenwärtige Situation in der Quantenmechanik (The present situation in quantum mechanics)*, Naturwissenschaften, Germany.

[Suzuki (2007)] Suzuki, T., H., Kaneda H., Nakagawa T., et al 2007,arXiv:0708.1829v1

[Theureau et al (1998)] Theureau, G., Bottinelli, L., Coudreau-Durand, N. et al. 1998, A&AS, 130, 333T

[Tully & Fisher (1977)] Tully, R. B. & Fisher, J. R. 1977, A&A, 54, 3, 661

[Tully et al (2008)] Tully, R. B. , Shaya E. J., Karachentsev I D., et al. 2008, ApJ, 676, 184T

[Vallejo et al (2002)] Vallejo, O., Braine, J., & Baudry, A. 2002, A&A, 387, 429

[Vollmer et al (1999)] Vollmer, B., Cayatte V., Boselli A., Balkowski C., Duschl W.J. 1999, A&A, 349, 411V

[Russell & Whitehead (1910-13)] Whitehead, Alfred North, and Russell, Bertrand, *Principia Mathematica*, 3 vols, Cambridge University Press, 1910, 1912, and 1913.

[Woods et al (1990)] Woods, D., Fahlman G.G., Madore B.F. 1990, ApJ, 353, 90

[Zwicky (1937)] Zwicky, F. 1937, ApJ, 86, 217Z

[Youman] Photo credit: Glen Youman, Penryn, California, www.astrophotos.net, by permission.

[Henley] Photo credit: J. R. Henley, Starfield Observatory, Nambour, Sunshine Coast, Queensland, Australia, by permission

[Hubble Space Telescope] Digital Photograph courtesy Digital Sky Survey, Hubble Space Telescope.

Internet Sources

[URL-1] Allen, Nick. *The Cepheid Distance Scale: A History*.

[URL-2] Barber Paradox, https://en.wikipedia.org/wiki/Barber_paradox

[URL-3] "The Nobel Prize in Physics 2011". Nobelprize.org. Retrieved 2011-10-06. https://WWW.nobelprize.org/prizes/physics/2011/summary/#

[URL-4] Mast Phase 2 (GSC2) Survey (2006), http://archive.stsci.edu/dss/

[URL-5] http://simbad.u-strasbg.fr/simbad/

[URL-6] Wikipedia contributors, *Occam's razor*, Wikipedia, The Free Encyclopedia, http://en.wikipedia.org/w/index.php? title=Occam

[URL-7] Wikipedia contributors, *Superman*, Wikipedia, The Free Encyclopedia, http://en.wikipedia.org/w/index.php? title=Superman&oldid=253177811, (accessed November 27, 2008).

[URL-8] Wikipedia contributors, *Gödel's incompleteness theorems*, Wikipedia, The Free Encyclopedia, http://en.wikipedia.org/w/index.php? title=G%C3%B6del%27s_incompleteness_theorems& oldid=256498126 (accessed December 27, 2008).

[URL-9] Wikipedia contributors, *Number*, Wikipedia, The Free Encyclopedia, http://en.wikipedia.org/w/index.php? title=Number&oldid=259687372 (accessed December 28, 2008).

[URL-10] Wikipedia contributors, *Schrödinger's cat,* Wikipedia, The Free Encyclopedia, http://en.wikipedia.org/w/index.php? title=Schr%C3%B6dinger%27s_cat&oldid=260423638 (accessed December 28, 2008).

[URL-11] Wikipedia contributors, *Electromagnetic tensor*, Wikipedia, The Free Encyclopedia, http://en.wikipedia.org/w/index.php? title=Electromagnetic_tensor&oldid=284404913 (accessed April 20, 2009).

[URL-12] Wikipedia contributors, Electromagnetic stress-energy tensor, Wikipedia, The Free Encyclopedia,
http://en.wikipedia.org/w/index.php?
title=Electromagnetic_stress-energy_tensor&oldid=284861941 (accessed April 20, 2009).

[URL-13] Peter M. Brown, *Faraday Tensor*,
http://www.geocities.com/physics_world /em/faraday_tensor.htm

[URL-14] Wikipedia contributors, *Alfred Adler*, Wikipedia, The Free Encyclopedia,
http://en.wikipedia.org/w/index.php?
title=Alfred_Adler&oldid=285172025 (accessed April 28, 2009).

[URL-15] Wikipedia contributors, *Pierre-Simon Laplace*, Wikipedia, The Free Encyclopedia,
http://en.wikipedia.org/w/index.php?
title=Pierre-Simon_Laplace&oldid=285641020 (accessed April 29, 2009).

[URL-16] Wikipedia contributors, *J·®zef Maria Hoene-Wro?ski*, Wikipedia, The Free Encyclopedia,
http://en.wikipedia.org/w/index.php?
title=J%C3%B3zef_Maria_Hoene-Wro%C5%84ski &oldid=281468392 (accessed April 30, 2009).

[URL-17] Wikipedia contributors, it Kerr metric, Wikipedia, The Free Encyclopedia,
http://en.wikipedia.org/w/index.php?
title=Kerr_metric&oldid=287433001 (accessed May 13, 2009).

www.ingramcontent.com/pod-product-compliance
Lightning Source LLC
Chambersburg PA
CBHW031848200326
41597CB00012B/317